10 Alaska AK STAR Grade 6 Math Practice Tests

The Ultimate Test Prep Collection with Answer Explanations

Dr. A. Nazari

10 Practice Tests

The Grand Championship Collection

Welcome, future Math Champion!

*You hold the **ultimate collection** —
ten full-length practice tests designed to take you
from first attempt to **complete mastery**.*

- *Conquer every Grade 6 topic*
- *Build unshakeable confidence*
- *Rise from Bronze to Gold to Champion*
- *Arrive at test day fully prepared*

The championship begins now.

*66 Ten tests may seem like
a marathon, but champi-
ons are made one step at a
time. Trust the process! 99*

The Champion's Path

Your 4-phase journey from Bronze to Champion

Bronze Round (Tests 1–3)

Your warm-up matches. Take these **untimed** to learn the format and set your baseline. Read the answer explanations after each test — this is where you build your foundation.

Silver Round (Tests 4–6)

Set a timer for **75 minutes**. Focus on the topics that tripped you up in Bronze. Practice showing your work on every problem. Your accuracy should be climbing.

Gold Round (Tests 7–9)

Full timed conditions (**60 minutes**). Simulate the real exam environment. Review only the questions you missed — targeted practice is the key to gold.

Championship Final (Test 10)

Your final match. Full exam conditions — timed, quiet, no breaks. This is your victory lap. Show yourself how far you've come!

Your Championship Kit

- **10 Full-Length Practice Tests** — every Grade 6 topic
- **Formula Reference Sheet**
- **Complete Answer Key** with explanations
- **Championship Scoreboard** to track your rise

Champion's Tip: Space your tests 2–3 days apart. Use the days in between for targeted review. By Test 10, you'll be amazed at your transformation.

♛ The Champion's Playbook ♛

I   **Read every question twice.** The first read tells you the topic. The second tells you exactly what to solve for. Champions never skim.

II   **Mark the clues.** Circle key numbers, underline the question, and cross out information that's just there to distract you.

III   **Choose your strategy.** Before touching pencil to paper, decide: Am I setting up a ratio? Solving an equation? Finding area? Name the approach.

IV   **Solve, then match.** For multiple choice — work the problem on scratch paper first, then find your answer among the choices.

V   **Eliminate and conquer.** Cross out obviously wrong answers. If you're left with two, you've already doubled your odds. Make an educated pick.

VI   **Estimate to verify.** After solving, ask: "Is this answer reasonable?" A quick mental estimate catches most calculation errors.

VII   **Leave nothing blank.** Even a well-reasoned guess is worth more than an empty space. Use partial work to support your answer.

🕐 Timing Mastery

Tests 1–3: **Untimed** (build foundation) > Tests 4–6: **75 min** (build speed) > Tests 7–10: **60 min** (championship conditions)

★ Grade 6 Championship Topics

🏅 Ratios & Proportions   🏅 Integers & Rational Numbers   🏅 Expressions & Equations

🏅 Geometry & Measurement   🏅 Statistics & Data Analysis

*A true champion isn't someone who never makes mistakes — it's someone who learns from **every single one**. After each test, review your errors carefully. That's where the real growth happens.*

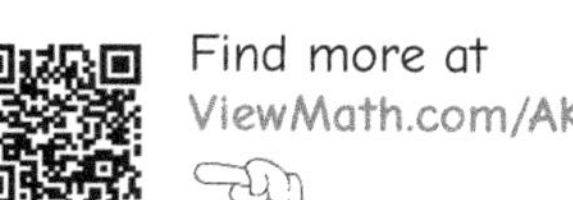

Find more at
ViewMath.com/AK-Grade6

The Champion's Toolkit

🏆 Required Equipment

Sharpened Pencils — Two #2 pencils — champions always have a backup

Quality Eraser — A clean, soft eraser that won't smudge your work

Scratch Paper — Blank paper for calculations, diagrams, and number lines

Ruler — Essential for geometry and coordinate plane questions

Timer — Begin using from the Silver Round onward

Quiet Workspace — A calm, well-lit area free from distractions

Permitted in Competition

- ✓ Pencil and eraser
- ✓ Scratch paper (provided)
- ✓ Ruler (if specified)
- ✓ Formula reference in this book

Not Permitted

- ✗ Calculators
- ✗ Electronic devices
- ✗ Textbooks or notes
- ✗ Outside help

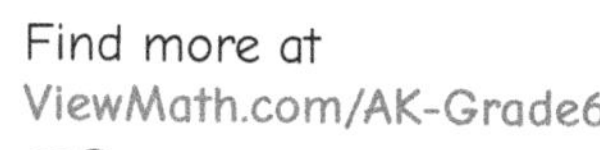

Find more at
ViewMath.com/AK-Grade6

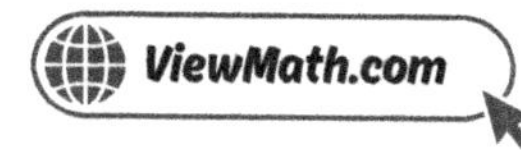

Formula Reference Sheet

🔺 Area Formulas

Rectangle	$A = l \times w$
Parallelogram	$A = b \times h$
Triangle	$A = \frac{1}{2} \times b \times h$
Trapezoid	$A = \frac{1}{2}(b_1 + b_2) \times h$

📦 Volume

Rectangular Prism $V = l \times w \times h$

🧊 Surface Area

Find the area of each face, then add them all up.

Rectangular Prism:

$SA = 2lw + 2lh + 2wh$

🔢 Order of Operations

P Parentheses first

E Exponents

M/D Multiply & Divide (left to right)

A/S Add & Subtract (left to right)

% Ratios & Percents

Ratio: $a : b$ or $\frac{a}{b}$

Unit rate: amount per 1 unit

Percent: a ratio out of 100

Part = Percent × Whole

⚖️ Integers & Absolute Value

Integers:

$\ldots, -3, -2, -1, 0, 1, 2, 3, \ldots$

$|-5| = 5$ $|5| = 5$

Absolute value = distance from 0

$\mathbf{X^1}$ Expressions & Equations

Exponent: $3^4 = 3 \times 3 \times 3 \times 3 = 81$

Variable: a letter that stands for a number

Equation: two expressions joined by $=$

Inequality: uses $<, >, \leq, \geq$

⊕ Coordinate Plane

Ordered pair: (x, y)

x-axis: horizontal y-axis: vertical

Origin: $(0, 0)$

Four quadrants (I, II, III, IV)

📊 Statistics

Mean: sum of values ÷ count

Median: middle value (sorted)

Range: max − min

Championship Scoreboard 

Track your rise through every championship round

Champion's Name: _______________________________

🏆 Round	🏅 Tier	📅 Date	⭐ Score	☺ Rating
1	Bronze			
2	Bronze			
3	Bronze			
4	Silver			
5	Silver			
6	Silver			
7	Gold			
8	Gold			
9	Gold			
10	👑			

♛ Champion's Reflection

My strongest topics (where I consistently score well):

____

____

Topics I improved on the most from Bronze to Gold:

____

____

My score trend (Bronze avg → Gold avg → Championship):

____

____

One strategy that helped me improve the most:

____

____

My confidence level for the real test (1–10): ________ / 10

Find more at
ViewMath.com/AK-Grade6

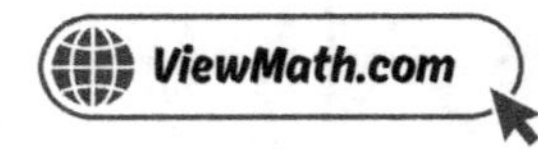

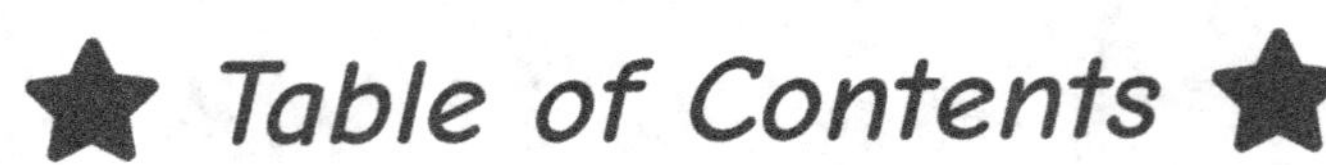 Table of Contents ⭐

Here's what we'll explore together!

 Let's learn and have fun!

Practice Test 1

📋 30 Questions

✏️ Before You Start ✏️

- ✔ **Read each question carefully** before choosing your answer.
- ✔ **Show your work** on scratch paper when you need to.
- ✔ **Skip hard questions** and come back to them later.
- ✔ **Check your answers** when you're done.
- ✔ **Take your time** — there's no rush!

⭐ You've Got This! ⭐

Do your best and show what you know!

1. A garden has flowers and vegetables in a ratio of 4 : 1. There are 20 flowers. How many vegetables are there?

Your Answer

2. The graph below shows the distance traveled by a delivery truck over time.

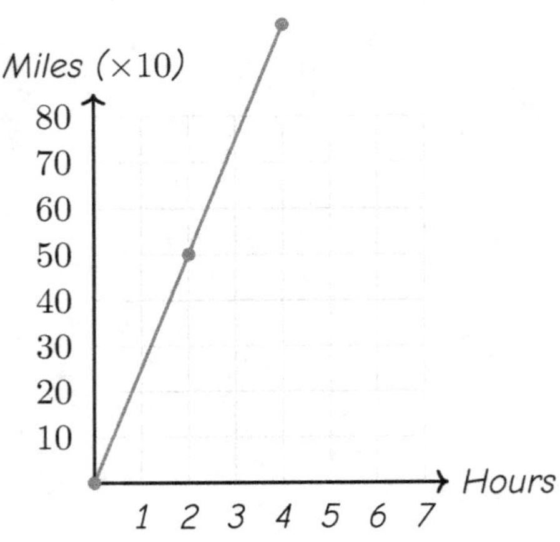

What is the truck's unit rate in miles per hour?

(A) 50 miles per hour

(B) 20 miles per hour

(C) 25 miles per hour

(D) 10 miles per hour

3. The double number line below shows equivalent ratios of scoops of mix to cups of water.

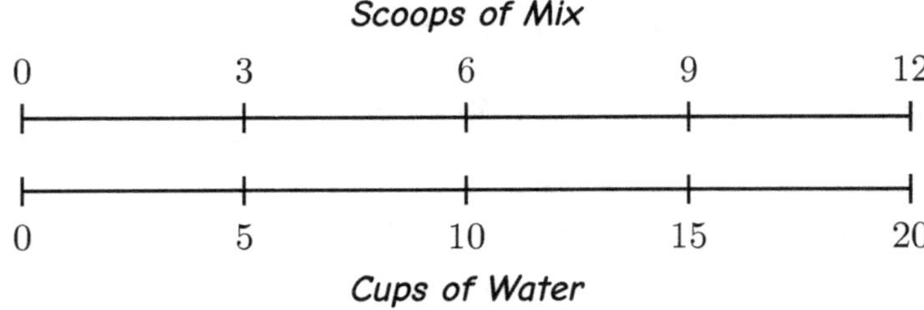

How many cups of water go with 9 scoops of mix?

(A) 10

(B) 12

(C) 15

(D) 18

Find more at
ViewMath.com/AK-Grade6

ViewMath.com

4. What is 72% written as a decimal?

(A) 72.0

(B) 7.2

(C) 0.072

(D) 0.72

5. The double number line below shows a map scale.

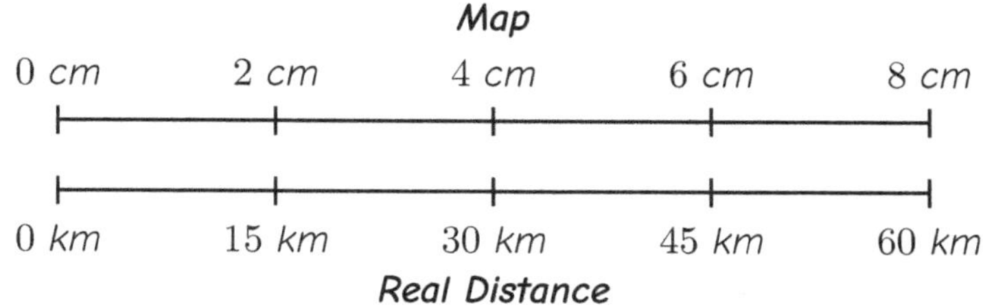

Two cities are 5 cm apart on the map. What is the real distance?

(A) 25 km

(B) 30 km

(C) 37.5 km

(D) 35 km

6. A carpenter has $2\frac{1}{2}$ feet of wood. Each shelf needs $\frac{5}{8}$ of a foot. How many shelves can the carpenter cut?

Your Answer:

7. The bar graph below shows the number of cans a school collected for recycling each week.

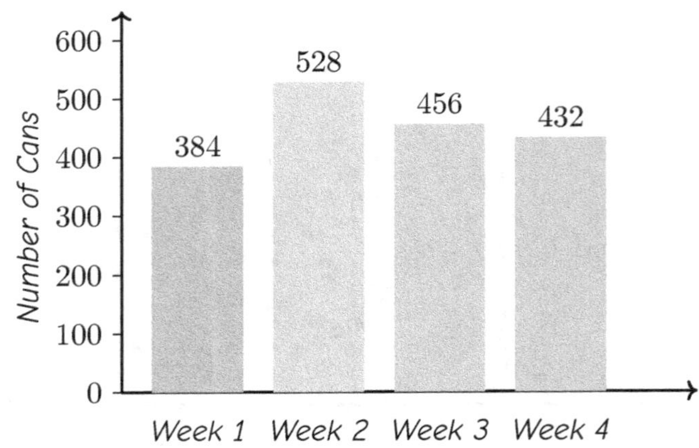

The school wants to divide all the cans equally among 12 classrooms. How many cans does each classroom receive?

(A) 180

(B) 1,800

(C) 150

(D) 130

8. A water jug holds 5.4 liters. How many liters are in 6 jugs?

(A) 32.4

(B) 324

(C) 3.24

(D) 30.4

9. Find $|-9| + |-6|$.

Your Answer:

10. The point $(-7, 3)$ is reflected across the x-axis. Write the coordinates of the reflected point.

Your Answer:

11. A store charges the prices shown below. Which expression gives the total cost for s sandwiches and d drinks?

Sandwich	Drink
$6 each	$2 each

(A) $6 + 2$

(B) $6s + 2d$

(C) $8sd$

(D) $2s + 6d$

12. A student made a factor tree for the expression $4(x + 3)$, shown below. Fill in the missing pieces labeled A and B.

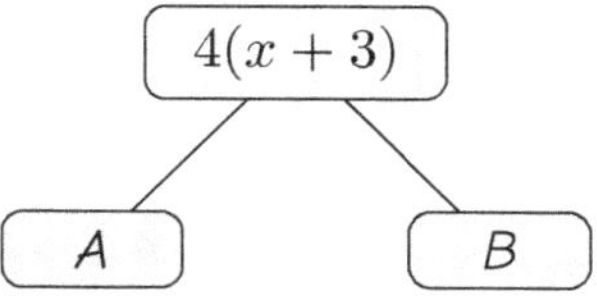

Your Answer

13. Evaluate $3(p + 2) - p$ when $p = 8$.

Your Answer

14. Which expression is equivalent to $3(x + 5)$?

(A) $3x + 5$

(B) $3x + 15$

(C) $8x$

(D) $3x + 8$

15. Write a situation that could be modeled by the expression $5n + 3$.

Your Answer

16. Solve: $\dfrac{m}{4} = 7$

(A) $m = 3$ (B) $m = 11$

(C) $m = 28$ (D) $m = 47$

17. Write a real-world situation that can be described by $x \le 15$.

Your Answer:

18. Is $x = 3$ a solution to $x \le 3$? Is $x = 3$ a solution to $x < 3$? Explain.

Your Answer:

19. Which table shows a relationship where y does NOT depend on x?

(A) $x : 1, 2, 3 \quad y : 3, 6, 9$ (B) $x : 1, 2, 3 \quad y : 7, 7, 7$

(C) $x : 1, 2, 3 \quad y : 2, 4, 6$ (D) $x : 1, 2, 3 \quad y : 5, 10, 15$

20. A triangular sail has a base of 3 m and a height of 7 m. How much fabric is needed to make 2 identical sails?

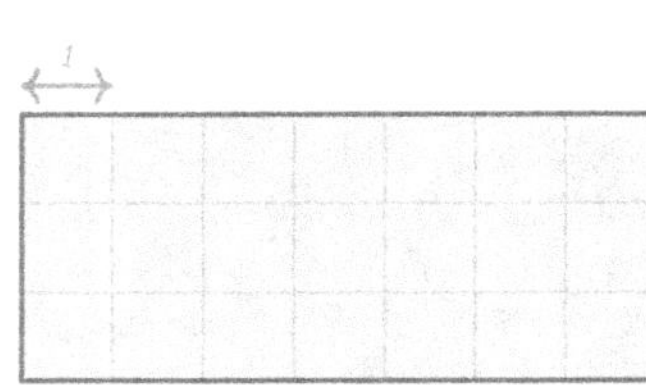

(A) $10.5 \ m^2$ (B) $42 \ m^2$

(C) $21 \ m^2$ (D) $20 \ m^2$

21. Points $A(0, 4)$ and $B(0, -5)$ are connected by a line segment. What is the length of $\overline{AB}$?

(A) 1 unit

(B) 5 units

(C) 4 units

(D) 9 units

22. A right triangle has vertices $(-4, 0)$, $(2, 0)$, and $(-4, 5)$. What is the area?

(A) 30 square units

(B) 15 square units

(C) 11 square units

(D) 20 square units

23. A rectangular prism has dimensions $10\ m$, $3\ m$, and $7\ m$. What is its surface area?

(A) $210\ m^2$

(B) $242\ m^2$

(C) $262\ m^2$

(D) $121\ m^2$

24. If you ask 20 classmates, "How many pets do you have?" and you get the answers $0, 1, 1, 2, 0, 3, 1, 2, 0, 1, 4, 2, 1, 0, 2, 1, 3, 1, 0, 2$, what does this tell you about the question?

(A) It is not a statistical question because some answers are 0.

(B) It is statistical because the answers vary.

(C) It is not statistical because the answers are whole numbers.

(D) It is statistical because there are 20 answers.

25. A dog sled team traveled $15, 18, 20, 22, 25$ miles on 5 days. What is the mean distance?

(A) 18

(B) 20

(C) 22

(D) 25

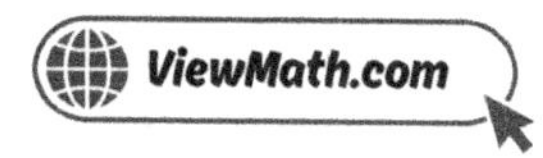

26. *The table below shows the distances (in miles) five students travel to school and their distances from the mean.*

Student	Miles	Distance from Mean
A	2	4
B	4	2
C	6	0
D	8	2
E	10	4

What is the MAD?

(A) 0

(B) 2

(C) 2.4

(D) 6

27. *A dot plot of students' shoe sizes shows a cluster from size 6 to size 8, with a gap at size 10 and one dot at size 12. What is the dot at size 12?*

(A) A cluster

(B) A gap

(C) A peak

(D) An outlier

28. *A box plot shows: min = 25, Q1 = 35, median = 50, Q3 = 65, max = 80. Find the range and IQR.*

Your Answer:

29. *School A has 150 sixth graders with a median reading score of 290 and IQR of 30. School B has 100 sixth graders with a median reading score of 285 and IQR of 45. Which school's students are more spread out in reading ability?*

(A) School A

(B) School B

(C) They are equally spread.

(D) Cannot be determined without the means.

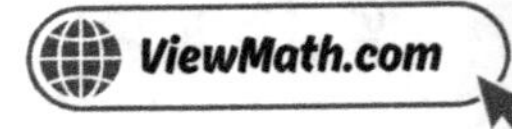

30. A line graph shows a town's average temperature each month from January (28°F) to May (62°F). The values increase each month. Is this graph showing a positive trend, a negative trend, or no trend?

Your Answer

 End of Practice Test 1 

Great job finishing the test!

 My Score

I got _____________ out of 30 questions right.

*Check your answers in the **Answer Key** at the back of the book.*

💡 *Review any questions you missed. That's how we learn!*

📊 Check Your Score Online!

Visit **ViewMath Academy** to enter your answers and see which topics you need to review. You can also explore lessons, take quizzes, track your scores, and save your progress!

viewmath.com/score/6.1.AK.16

Or go to viewmath.com/score and enter code: 6.1.AK.16

2

Practice Test 2

☑ 30 Questions

✏ Before You Start ✏

- ✓ **Read each question carefully** before choosing your answer.
- ✓ **Show your work** on scratch paper when you need to.
- ✓ **Skip hard questions** and come back to them later.
- ✓ **Check your answers** when you're done.
- ✓ **Take your time** — there's no rush!

★ You've Got This! ★

Do your best and show what you know!

1. *Look at the tape diagram below.*

Boys: ▨▨▨▨
Girls: ☐☐☐☐☐☐

There are 24 girls. How many boys are there?

(A) 4

(B) 12

(C) 16

(D) 20

2. *The table shows the earnings of two babysitters.*

	Hours Worked	Earnings
Lily	5	$60
Noah	4	$52

Part A: *Find each babysitter's unit rate (dollars per hour).*

Part B: *Who earns more per hour? How much more?*

Your Answer:

Find more at
ViewMath.com/AK-Grade6

3. The graph below shows points that represent equivalent ratios of flour to sugar in a recipe.

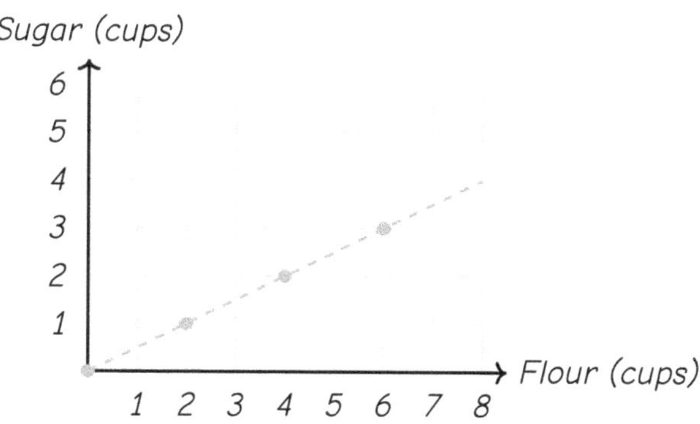

Part A: What is the ratio of flour to sugar?

Part B: If you need 8 cups of flour, how many cups of sugar do you need?

Your Answer:

4. Which of the following is equal to 50%?

(A) $\frac{1}{5}$

(B) 0.05

(C) $\frac{1}{2}$

(D) 5.0

5. A pool fills at a rate of 12 gallons per minute. How long does it take to fill 360 gallons?

(A) 25 minutes

(B) 30 minutes

(C) 36 minutes

(D) 40 minutes

6. Maya jogged $\frac{7}{8}$ of a mile. Each lap around the park is $\frac{1}{4}$ of a mile. How many laps did she complete?

(A) $\frac{7}{32}$

(B) 4

(C) 2

(D) $3\frac{1}{2}$

7. *What is $756 \div 3$?*

(A) 252

(B) 250

(C) 258

(D) 202

8. *A book costs $12.95 and a bookmark costs $3.48. Jenna buys one of each. How much does she spend in all?*

Your Answer:

9. *What is $|-12|$?*

(A) -12

(B) 12

(C) 0

(D) $-(-(-12))$

10. *Explain why the point $(9,0)$ is not in any quadrant.*

Your Answer:

11. *A baker makes c cupcakes. She packs them equally into 6 boxes. Which expression shows the number in each box?*

(A) $c-6$

(B) $6c$

(C) $c \div 6$

(D) $c+6$

12. *In the expression $5(y+8)$, name the two factors.*

Your Answer:

Find more at
ViewMath.com/AK-Grade6

ViewMath.com

13. The triangle below has a base of $b = 10$ cm and a height of $h = 6$ cm. Use the formula $A = \frac{1}{2} \times b \times h$ to find its area.

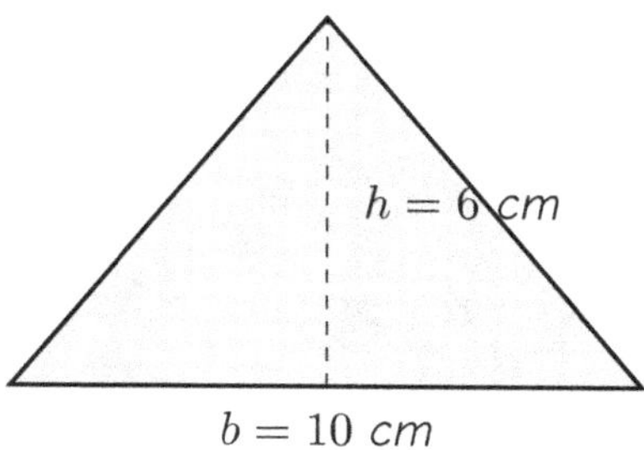

Your Answer

14. Simplify: $3a + 7 + 5a - 2a + 3$

Your Answer

15. A movie ticket costs \$9. Your group also shares a \$6 bucket of popcorn. Which expression represents the total cost for p people?

(A) $9p + 6$

(B) $9 + 6p$

(C) $15p$

(D) $9p - 6$

16. Solve: $k + 15 = 32$

(A) $k = 47$

(B) $k = 17$

(C) $k = 27$

(D) $k = 2$

17. Which inequality represents "a number y is fewer than 20"?

(A) $y > 20$

(B) $y \geq 20$

(C) $y < 20$

(D) $y \leq 20$

18. Which inequality has a graph with an open circle at 0 and shading to the right?

(A) $x \leq 0$

(B) $x \geq 0$

(C) $x > 0$

(D) $x < 0$

19. Identify the independent and dependent variables: A runner jogs 6 miles per hour. The distance d depends on how many hours h the runner jogs.

Your Answer

20. A triangle has base 14 m and height 8 m. What is its area?

(A) $112\ m^2$

(B) $22\ m^2$

(C) $44\ m^2$

(D) $56\ m^2$

21. A rectangle on the coordinate plane has a length of 10 units and width of 4 units. One vertex is at the origin $(0,0)$ and the sides are along the axes. Which of the following could be the opposite vertex?

(A) $(10, 4)$

(B) $(14, 0)$

(C) $(4, 4)$

(D) $(10, 10)$

22. A square on the coordinate plane has vertices $(-3, -3)$, $(5, -3)$, $(5, 5)$, and $(-3, 5)$. What is the area?

Your Answer

Find more at
ViewMath.com/AK-Grade6

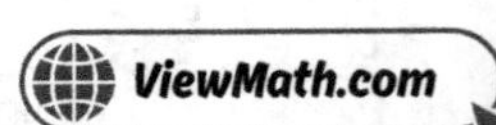

23. How many faces does a rectangular prism have?

(A) 4

(B) 5

(C) 6

(D) 8

24. A student says, "Any question with a number for an answer is statistical." Is the student correct? Explain.

Your Answer:

25. The bar graph below shows average monthly temperatures (°F) in Fairbanks for four months.

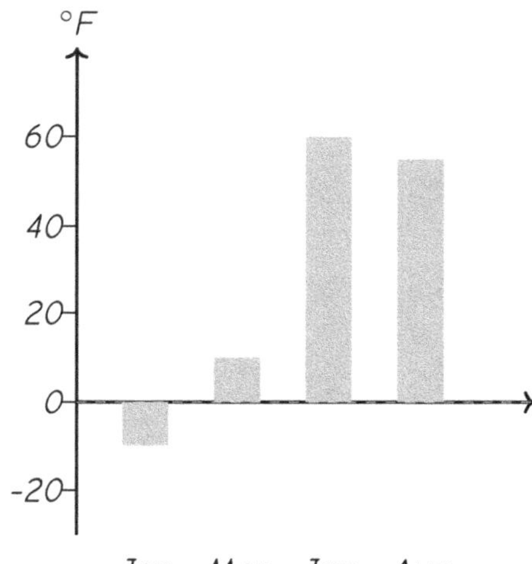

The temperatures are approximately −10°F, 10°F, 60°F, 55°F. What is the mean of these four monthly temperatures?

(A) 20.5°F

(B) 28.75°F

(C) 32.5°F

(D) 55°F

26. The dot plot below shows the number of laps students ran during gym class.

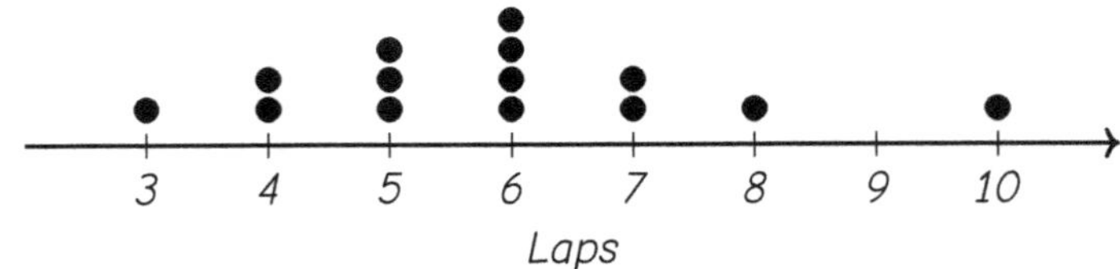

Find the range and IQR from the dot plot data.

Your Answer:

27. Which type of data display shows every individual data value?

(A) Histogram

(B) Circle graph

(C) Dot plot

(D) Box plot

28. Which data set has the five-number summary: $5, 10, 15, 20, 25$?

(A) $5, 10, 15, 20, 25$

(B) $5, 8, 10, 15, 18, 20, 22, 25$

(C) $5, 7, 10, 12, 15, 18, 20, 23, 25$

(D) Any of these could have that five-number summary.

29. A dot plot shows test scores for Class 1: most dots clustered between 75 and 85 with one dot at 40. Class 2's dot plot shows dots evenly spread from 55 to 95. Which class is likely more consistent?

(A) Class 1, because most scores are close together.

(B) Class 2, because the scores are evenly spread.

(C) Neither — they are equally consistent.

(D) Cannot be determined without the means.

Find more at
ViewMath.com/AK-Grade6

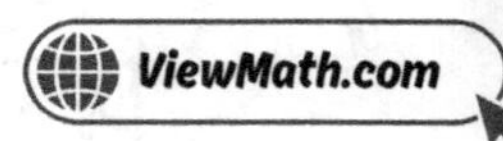

30. *The double bar graph below shows the number of books read by boys and girls in two classes. Use the graph to answer the question.*

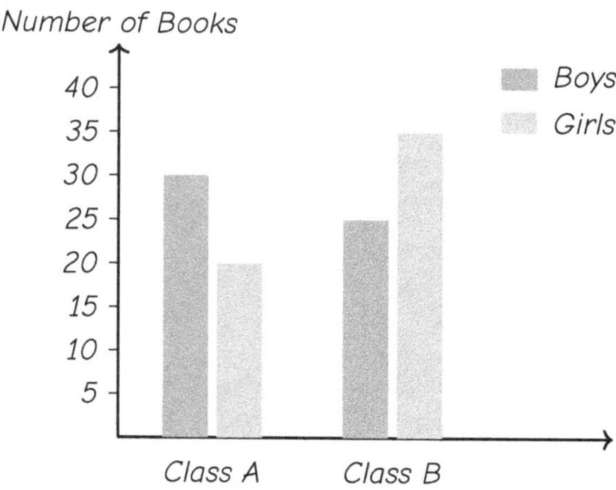

How many more books did boys in Class A read than girls in Class A?

(A) 5

(B) 10

(C) 15

(D) 20

⭐ End of Practice Test 2 ⭐

Great job finishing the test!

 My Score

I got ___________ out of 30 questions right.

*Check your answers in the **Answer Key** at the back of the book.*

💡 *Review any questions you missed. That's how we learn!*

📊 Check Your Score Online!

Visit **ViewMath Academy** to enter your answers and see which topics you need to review. You can also explore lessons, take quizzes, track your scores, and save your progress!

viewmath.com/score/6.1.AK.17

Or go to viewmath.com/score and enter code: 6.1.AK.17

3

Practice Test 3

☑ 30 Questions

✏ Before You Start ✏

- ✔ **Read each question carefully** before choosing your answer.
- ✔ **Show your work** on scratch paper when you need to.
- ✔ **Skip hard questions** and come back to them later.
- ✔ **Check your answers** when you're done.
- ✔ **Take your time** — there's no rush!

★ You've Got This! ★

Do your best and show what you know!

1. A baker uses 2 eggs **per** 3 cups of flour. Which ratio represents flour to eggs?

   (A) 2 : 3   (B) 3 : 5

   (C) 3 : 2   (D) 2 : 5

2. A box of 12 muffins costs \$9. What is the cost per muffin?

   (A) \$1.25   (B) \$0.75

   (C) \$1.33   (D) \$3.00

3. A store sells 7 bananas for \$2. How much do 21 bananas cost?

   (A) \$4   (B) \$6

   (C) \$9   (D) \$14

4. A store has 200 items. 50% are on sale. How many items are on sale?

   (A) 50   (B) 150

   (C) 100   (D) 200

5. A store sells 3 T-shirts for \$24. How much do 10 T-shirts cost?

   (A) \$72   (B) \$80

   (C) \$70   (D) \$90

6. What is $\dfrac{3}{5} \div \dfrac{1}{5}$?

   (A) $\dfrac{3}{25}$ (B) 3

   (C) $\dfrac{1}{3}$ (D) 15

Find more at
ViewMath.com/AK-Grade6

ViewMath.com

7. *What is* $7{,}344 \div 24$*?*

    (A) 36   (B) 360

    (C) 306   (D) 3,006

8. *What is* 1.2×0.3*?*

    (A) 3.6   (B) 0.36

    (C) 36   (D) 0.036

9. *Which number has an absolute value of 6?*

    (A) *6 only*   (B) *−6 only*

    (C) *Both 6 and −6*   (D) *No number has an absolute value of 6*

10. *Which sign convention describes all points in Quadrant II?*

    (A) $(+,+)$ (B) $(-,+)$

    (C) $(-,-)$ (D) $(+,-)$

11. *Sam has* x *stickers. He gives away 6. Which expression shows how many stickers Sam has now?*

    (A) $x + 6$ (B) $6x$

    (C) $6 - x$ (D) $x - 6$

12. *How many terms does the expression* $4(a + b)$ *have when written in its factored form?*

Your Answer:

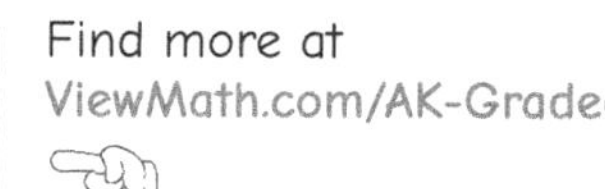

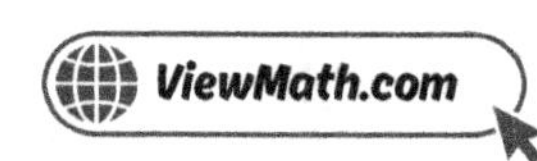

13. Evaluate $k^2 + k$ when $k = 3$.

(A) 6

(B) 9

(C) 12

(D) 15

14. Use the distributive property to expand $8(y + 3)$.

Your Answer

15. In the formula $d = 60t$, what could d and t represent?

(A) d = days, t = hours

(B) d = distance in miles, t = time in hours

(C) d = dollars, t = tax rate

(D) d = degrees, t = temperature

16. Solve: $x + 24 = 50$

(A) $x = 26$

(B) $x = 74$

(C) $x = 36$

(D) $x = 2$

17. The table below matches phrases to inequality symbols. Which row has an error?

Row	Phrase	Symbol
1	At least	$\geq$
2	No more than	$\leq$
3	Fewer than	$<$
4	At most	$<$

(A) Row 1

(B) Row 2

(C) Row 3

(D) Row 4

Find more at
ViewMath.com/AK-Grade6

ViewMath.com

18. The pool opens when the temperature is more than 75°F. Which describes the graph of this inequality?

(A) Closed circle at 75, shade right

(B) Open circle at 75, shade right

(C) Closed circle at 75, shade left

(D) Open circle at 75, shade left

19. A recipe uses 2 cups of flour for every batch of cookies. If you make b batches, which equation gives the total flour f?

(A) $f = b + 2$

(B) $f = 2b$

(C) $b = 2f$

(D) $f = b \div 2$

20. A triangular pennant has a base of 5 in and a height of 12 in. What is its area?

(A) $60\ in^2$

(B) $17\ in^2$

(C) $30\ in^2$

(D) $34\ in^2$

21. What is the distance between $(-6, -2)$ and $(-6, 5)$?

(A) 3 units

(B) 7 units

(C) 12 units

(D) 6 units

22. Find the area of the shaded triangle on the coordinate plane below.

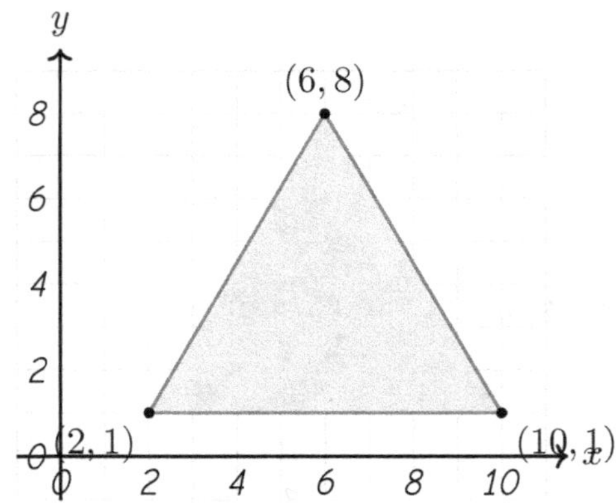

Your Answer:

23. A rectangular prism has a surface area of 94 cm^2. Its length is 5 cm, width is 3 cm, and height is 4 cm. Is this correct?

(A) Yes, the surface area is 94 cm^2.

(B) No, the surface area is 120 cm^2.

(C) No, the surface area is 60 cm^2.

(D) No, the surface area is 47 cm^2.

24. Which question would produce data where every answer is the same?

(A) How many hours do you watch TV each day?

(B) How many feet are in one yard?

(C) What is your favorite sport?

(D) How far can you throw a ball?

25. Five Anchorage students tracked miles hiked: $3, 5, 4, 6, 22$. A sixth student hiked 5 miles. How does adding this value affect the mean?

(A) The mean increases because 5 is a high value.

(B) The mean decreases because 5 is below the current mean of 8.

(C) The mean stays the same.

(D) The median changes but the mean does not.

Find more at
ViewMath.com/AK-Grade6

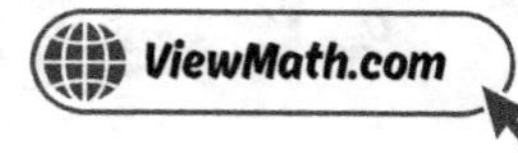

26. *Data:* $50, 55, 60, 65, 70.$ *The mean is 60. Find the MAD.*

Your Answer

27. A dot plot and a histogram are made from the same data set of 12 values. Which statement is true?

(A) The dot plot shows exactly 12 dots.

(B) The histogram shows exactly 12 bars.

(C) Both must have the same number of bars and dots.

(D) The dot plot always has more dots than the histogram has bars.

28. A box plot has min $= 10$, $Q1 = 20$, median $= 30$, $Q3 = 40$, max $= 50$. What is the IQR?

(A) 10

(B) 20

(C) 30

(D) 40

29. Heights (inches) of Team 1: $60, 62, 63, 64, 65, 66, 68.$ Heights of Team 2: $58, 60, 61, 64, 67, 69, 72.$ Which team has the greater range?

(A) Team 1

(B) Team 2

(C) They are equal.

(D) Cannot be determined.

30. A frequency table records the number of pets owned by 25 students. The frequencies are: 0 pets (6), 1 pet (10), 2 pets (5), 3 pets (4). How many students own at least 2 pets?

(A) 5

(B) 9

(C) 4

(D) 15

 End of Practice Test 3 

Great job finishing the test!

 My Score

I got _____________ out of 30 questions right.

Check your answers in the **Answer Key** at the back of the book.

 Review any questions you missed. That's how we learn!

Check Your Score Online!

Visit **ViewMath Academy** to enter your answers and see which topics you need to review. You can also explore lessons, take quizzes, track your scores, and save your progress!

viewmath.com/score/6.1.AK.18

Or go to viewmath.com/score and enter code: 6.1.AK.18

Practice Test 4

 30 Questions

✏ Before You Start ✏

- ✓ **Read each question carefully** before choosing your answer.
- ✓ **Show your work** on scratch paper when you need to.
- ✓ **Skip hard questions** and come back to them later.
- ✓ **Check your answers** when you're done.
- ✓ **Take your time** — there's no rush!

 You've Got This! 

Do your best and show what you know!

1. A store sells 3 pencils **for every** 1 eraser. Which ratio represents pencils to erasers?

   (A) $1:3$ (B) $3:1$

   (C) $3:4$ (D) $1:4$

2. A bike travels at a unit rate of 14 miles per hour. How long will it take to travel 49 miles?

   Your Answer:

3. The ratio table below has an error in one row. Which row has the error?

x	y
4	10
8	20
12	25
16	40

   (A) Row 1   (B) Row 2

   (C) Row 3   (D) Row 4

4. Convert 0.875 to a percent and then to a fraction in simplest form.

   Your Answer:

5. A car travels 180 miles in 3 hours. At the same rate, how far will it go in 7 hours?

   (A) 360 miles   (B) 420 miles

   (C) 540 miles   (D) 300 miles

Find more at
ViewMath.com/AK-Grade6

6. What is $\dfrac{2}{3} \div \dfrac{4}{5}$?

(A) $\dfrac{8}{15}$ (B) $\dfrac{6}{5}$

(C) $\dfrac{5}{6}$ (D) $\dfrac{2}{15}$

7. In the standard long-division algorithm, the four repeating steps in order are: Divide, ________, Subtract, Bring Down. What is the missing step?

(A) Estimate (B) Multiply

(C) Check (D) Add

8. Compute 3.6×0.4.

Your Answer

9. A diver is at -40 feet and a bird is at 25 feet above sea level. Who is farther from sea level?

(A) The bird, because $25 > -40$ (B) They are the same distance from sea level

(C) The diver, because $|-40| > |25|$ (D) Neither, because you cannot compare positive and negative numbers

10. In which quadrant is the point $(-4, 7)$ located?

(A) Quadrant I (B) Quadrant II

(C) Quadrant III (D) Quadrant IV

Find more at
ViewMath.com/AK-Grade6

ViewMath.com

11. Which phrase matches the expression $\frac{n}{3} + 7$?

(A) 7 more than a number n divided by 3

(B) 3 divided by n plus 7

(C) 7 times n divided by 3

(D) n plus 7, divided by 3

12. Identify the constant(s) in the expression $3m + 7 - 2m + 4$.

Your Answer:

13. Look at the formula and the rectangle below. What is the perimeter of the rectangle?

$$w = 3\ cm$$

$$P = 2l + 2w$$

$$l = 7\ cm$$

(A) 10 cm

(B) 20 cm

(C) 21 cm

(D) 42 cm

14. Simplify: $2(4m + 1) + 3m$

Your Answer:

15. In the formula $P = 4s$, the variable s represents the side length of a square. What does P represent?

(A) The area of the square

(B) The perimeter of the square

(C) The number of sides

(D) The diagonal of the square

16. The number line below shows a jump. Which equation does this model represent?

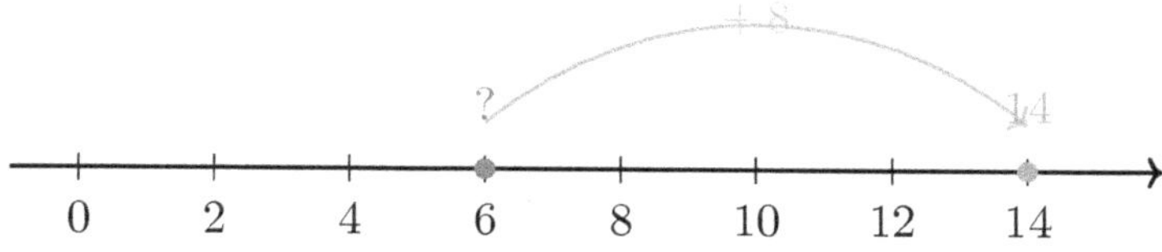

(A) $x + 8 = 14$, so $x = 6$

(B) $x - 8 = 14$, so $x = 22$

(C) $8x = 14$, so $x = 1.75$

(D) $x + 14 = 8$, so $x = -6$

17. Which of the following values is NOT a solution to $y \geq 3$?

(A) $y = 3$

(B) $y = 4$

(C) $y = 100$

(D) $y = 2.5$

18. Describe the graph of $n \leq 5$.

(A) Open circle at 5, shade left

(B) Closed circle at 5, shade left

(C) Open circle at 5, shade right

(D) Closed circle at 5, shade right

19. The number of inches i equals 12 times the number of feet f: $i = 12f$. How many inches are in 5 feet?

(A) 17 inches

(B) 48 inches

(C) 60 inches

(D) 125 inches

20. A triangle has base 20 in and height 7 in. What is the area?

(A) 140 in^2

(B) 27 in^2

(C) 70 in^2

(D) 54 in^2

21. A rectangle has vertices $(1, 2)$, $(7, 2)$, $(7, -4)$, and $(1, -4)$. What is the length of the longer side?

(A) 5 units

(B) 6 units

(C) 7 units

(D) 8 units

22. A rectangle has vertices $(-1, 3)$, $(5, 3)$, $(5, -2)$, and $(-1, -2)$. What is the area?

(A) 25 square units

(B) 30 square units

(C) 22 square units

(D) 35 square units

23. Which net below could fold into a cube?

A   B   C

(A) Net A only

(B) Net B only

(C) Net C only

(D) Nets A and B

24. Explain why "What is my teacher's age?" is **not** a statistical question.

Your Answer:

25. A wildlife survey counted $6, 9, 7, 11, 12$ moose in 5 areas. What is the median?

(A) 7

(B) 9

(C) 11

(D) 12

Find more at
ViewMath.com/AK-Grade6

ViewMath.com

26. Data A: IQR = 4. Data B: IQR = 20. What can you conclude?

(A) Data A has a higher median.

(B) Data B has a higher median.

(C) The middle 50% of Data A is more tightly clustered than Data B.

(D) Data A has more data points.

27. Why are there no gaps between the bars in a histogram?

(A) Because histograms always have equal bars.

(B) Because the intervals are consecutive and cover all values with no numbers left out.

(C) Because gaps would make it look like a dot plot.

(D) Because all data values are the same.

28. A box plot has a very long right whisker and a short left whisker. What does this suggest?

(A) The data is symmetric.

(B) The data is skewed right (has high outliers or a long tail to the right).

(C) The data is skewed left.

(D) All values are the same.

29. Two box plots are shown for heights (in cm) of plants grown with Fertilizer A and Fertilizer B.

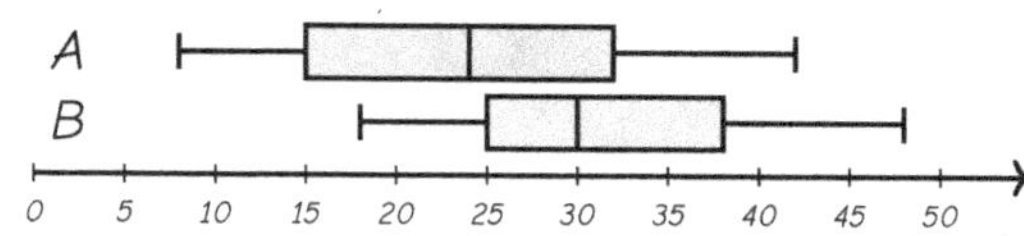

Write the five-number summary for each fertilizer. Compare the medians, ranges, and IQRs. Which fertilizer produced taller plants on average? Which produced more consistent growth?

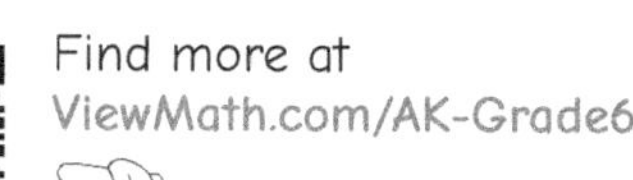

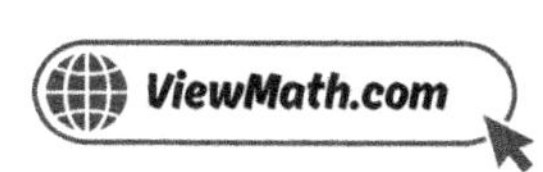

30. *On a line graph, the line goes steeply upward from April to May and then stays flat from May to June. What does this tell you?*

(A) The value decreased from April to June

(B) The value increased sharply, then stayed the same

(C) The value stayed the same the entire time

(D) The value increased at a steady rate

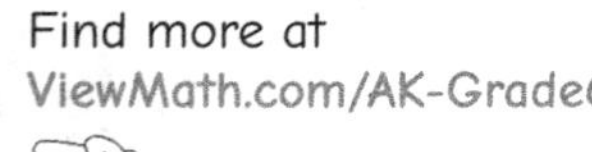

Find more at
ViewMath.com/AK-Grade6

 # End of Practice Test 4 

Great job finishing the test!

 ## My Score

I got ____________ out of 30 questions right.

Check your answers in the **Answer Key** at the back of the book.

Review any questions you missed. That's how we learn!

📊 Check Your Score Online!

Visit **ViewMath Academy** to enter your answers and see which topics you need to review. You can also explore lessons, take quizzes, track your scores, and save your progress!

viewmath.com/score/6.1.AK.19

Or go to viewmath.com/score and enter code: 6.1.AK.19

Practice Test 5

 30 Questions

 Before You Start 

- ✓ **Read each question carefully** before choosing your answer.
- ✓ **Show your work** on scratch paper when you need to.
- ✓ **Skip hard questions** and come back to them later.
- ✓ **Check your answers** when you're done.
- ✓ **Take your time** — there's no rush!

 You've Got This! 

Do your best and show what you know!

1. A class votes on two activities: hiking and swimming. The ratio of votes for hiking to swimming is $3 : 2$. There are 25 votes total. How many voted for swimming?

Your Answer:

2. A car travels 240 miles using 8 gallons of gas. What is the unit rate in miles per gallon?

(A) 8

(B) 32

(C) 30

(D) 248

3. It takes 6 cups of flour to make 4 loaves of bread. How many cups of flour for 10 loaves?

(A) 12

(B) 10

(C) 15

(D) 20

4. The bar model below shows the results of a class vote on a field trip destination.

| Zoo (60%) | Farm | Park |

Total: 100%

If the Farm section is twice as wide as the Park section, what percent voted for the Farm?

(A) 10%

(B) 40%

(C) $\dfrac{20}{3}\%$

(D) $\approx 27\%$

5. A map scale says $1\ cm = 5\ km$. Two cities are 8 cm apart on the map. What is the real distance?

(A) 13 km

(B) 5 km

(C) 40 km

(D) 80 km

Find more at
ViewMath.com/AK-Grade6

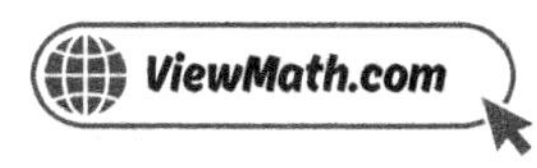

6. A board is $\frac{7}{8}$ of a meter long. You need pieces that are $\frac{1}{4}$ of a meter. How many pieces can you cut? Express your answer as a fraction or mixed number.

Your Answer:

7. What is $1{,}575 \div 5$?

(A) 305

(B) 3,150

(C) 315

(D) 351

8. When you multiply 0.6×0.7, how many decimal places should the product have?

(A) 0

(B) 1

(C) 2

(D) 3

9. Which statement is **true**?

(A) $|-10| = -10$

(B) $|-10| = 10$

(C) $|-10| = 0$

(D) $|-10| = -(-(-10))$

10. Starting at the origin, you move 4 units to the left and 3 units up, then 7 units to the right and 5 units down. What point do you end at?

Your Answer:

11. Write an expression for: "9 fewer than a number z."

Your Answer:

Find more at
ViewMath.com/AK-Grade6

ViewMath.com

12. In the expression $6(n + 2)$, what are the two factors?

   (A) 6 and n

   (B) 6 and 2

   (C) n and 2

   (D) 6 and $(n + 2)$

13. Evaluate $m^2 + 2m + 1$ when $m = 4$

   Your Answer

14. Simplify: $4(3y - 2)$

   (A) $12y - 2$

   (B) $12y - 8$

   (C) $7y - 2$

   (D) $7y - 6$

15. A store sells notebooks for $3 each and pens for $1 each. Which expression gives the total cost for n notebooks and p pens?

   (A) $3 + n + 1 + p$

   (B) $3n + p$

   (C) $4np$

   (D) $n + 3p$

16. Solve: $\dfrac{w}{8} = 9$

   Your Answer

17. Which phrase matches the inequality $t \leq 30$?

   (A) The temperature is more than 30 degrees

   (B) The temperature is exactly 30 degrees

   (C) The temperature is no more than 30 degrees

   (D) The temperature is at least 30 degrees

18. Which inequality matches a graph with a closed circle at 3 and shading to the right?

(A) $x > 3$

(B) $x < 3$

(C) $x \geq 3$

(D) $x \leq 3$

19. In the equation $y = x + 4$, which statement is true?

(A) x is the dependent variable

(B) y is always 4

(C) x is the independent variable

(D) y decreases as x increases

20. Two triangles both have a base of 10 cm. Triangle P has a height of 6 cm and Triangle Q has a height of 8 cm. How much greater is the area of Triangle Q?

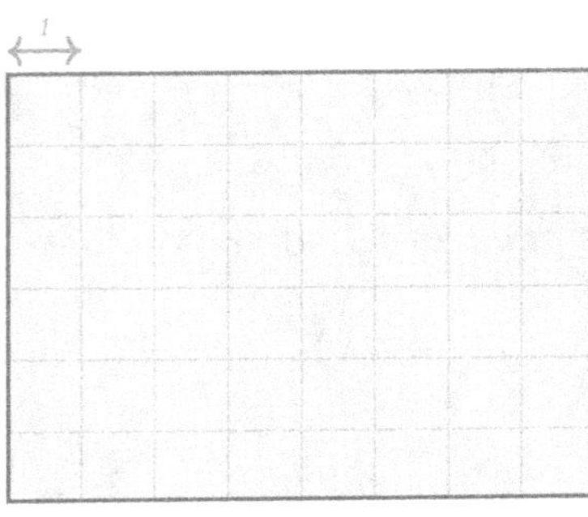

(A) $2\ cm^2$

(B) $10\ cm^2$

(C) $20\ cm^2$

(D) $40\ cm^2$

21. A square on the coordinate plane has vertices $(0, 0)$, $(0, 5)$, $(5, 5)$, and $(5, 0)$. What is the perimeter?

Your Answer:

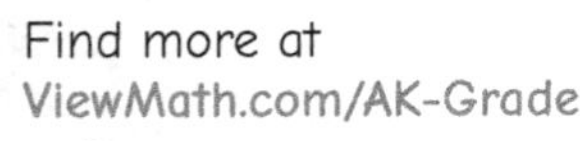
Find more at
ViewMath.com/AK-Grade6

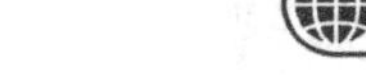

22. *A right triangle has vertices* $(2, 1)$, $(2, 7)$, *and* $(8, 1)$. *What is the area?*

(A) 36 *square units*

(B) 18 *square units*

(C) 12 *square units*

(D) 24 *square units*

23. *A box is* 5 *in long,* 5 *in wide, and* 10 *in tall. What is the surface area?*

(A) 250 *in*^{2}

(B) 200 *in*^{2}

(C) 300 *in*^{2}

(D) 150 *in*^{2}

24. *Which of the following is a statistical question?*

(A) *How many days are in February during a leap year?*

(B) *How many hours do sixth graders spend on homework each week?*

(C) *What is the boiling point of water in degrees Celsius?*

(D) *How many states are in the United States?*

25. *A trapper recorded catching* 4, 6, 5, 7, 3 *animals over* 5 *days. What is the mean?*

(A) 4

(B) 5

(C) 6

(D) 7

26. *Data:* 20, 22, 24, 26, 28. *The mean is* 24. *What is the MAD?*

(A) 2

(B) 2.4

(C) 4

(D) 8

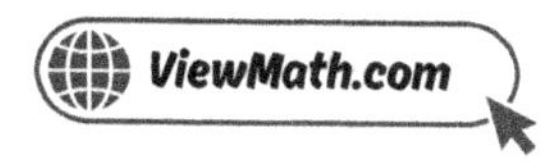

27. A histogram shows homework times: 0–14 min (5 students), 15–29 min (10 students), 30–44 min (8 students), 45–59 min (2 students). How many students are represented?

> Your Answer:

28. A box plot has the median line closer to Q1 than to Q3. What does this mean?

(A) The data is perfectly symmetric.

(B) The upper half of the data is more spread out than the lower half.

(C) The lower half is more spread out.

(D) The data has no outliers.

29. Weekly allowances of two groups of students are listed below.

Group X: $5, $8, $10, $10, $12, $15, $20

Group Y: $9, $10, $10, $11, $11, $12, $12

Find the median, range, and IQR for each group. Write 2–3 sentences comparing the groups.

> Your Answer:

30. A bar graph is missing its title. The horizontal axis is labeled "Favorite Sport" and the vertical axis is labeled "Number of Students." Which title best fits this graph?

(A) Students' Heights in Sixth Grade

(B) Favorite Sports of Sixth Graders

(C) Monthly Rainfall in Our City

(D) Temperature Over One Week

 End of Practice Test 5 

Great job finishing the test!

 My Score

I got _____________ out of 30 questions right.

Check your answers in the Answer Key at the back of the book.

 Review any questions you missed. That's how we learn!

📊 Check Your Score Online!

Visit **ViewMath Academy** to enter your answers and see which topics you need to review. You can also explore lessons, take quizzes, track your scores, and save your progress!

viewmath.com/score/6.1.AK.20

Or go to viewmath.com/score and enter code: 6.1.AK.20

Practice Test 6

 30 Questions

✏️ **Before You Start** ✏️

- ✓ **Read each question carefully** before choosing your answer.
- ✓ **Show your work** on scratch paper when you need to.
- ✓ **Skip hard questions** and come back to them later.
- ✓ **Check your answers** when you're done.
- ✓ **Take your time** — there's no rush!

 You've Got This! 

Do your best and show what you know!

1. A smoothie recipe uses 3 bananas **for every** 4 cups of yogurt. Write the ratio of bananas to yogurt. Then write the ratio of yogurt to bananas.

> Your Answer

2. Brand X sells 5 notebooks for $8.75. Brand Y sells 3 notebooks for $4.50. Which is the better deal?

- (A) Brand X at $1.75 each
- (B) Brand Y at $1.50 each
- (C) Brand X at $1.50 each
- (D) Brand Y at $1.75 each

3. A teacher hands out 3 pencils for every 2 students. There are 16 students. How many pencils does she need?

- (A) 18
- (B) 24
- (C) 20
- (D) 32

4. What is 0.35 written as a percent?

- (A) 0.35%
- (B) 3.5%
- (C) 35%
- (D) 350%

5. Which strategy is best for solving "how much for one?" problems?

- (A) Tape diagram
- (B) Unit rate
- (C) Ratio table
- (D) Equation

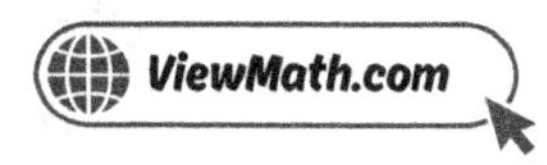

6. What is the reciprocal of $\frac{3}{7}$?

   (A) $\frac{3}{7}$

   (C) $-\frac{3}{7}$

   (B) $\frac{7}{3}$

   (D) $\frac{1}{3}$

7. A baker makes 2,016 cookies and packs 14 cookies per bag. How many bags can she fill?

   (A) 142

   (C) 148

   (B) 144

   (D) 104

8. What is $9.36 \div 1.2$?

   (A) 78

   (C) 7.8

   (B) 0.78

   (D) 7.08

9. What is the opposite of -3?

   (A) -3

   (C) $\frac{1}{3}$

   (B) 0

   (D) 3

10. Starting at the origin, how do you plot the point $(-3, 4)$?

   (A) Move 3 units right and 4 units up

   (C) Move 3 units left and 4 units up

   (B) Move 3 units left and 4 units down

   (D) Move 4 units left and 3 units up

11. Which expression represents "6 more than half of a number n"?

(A) $6 + 2n$

(B) $\dfrac{n+6}{2}$

(C) $\dfrac{n}{2} + 6$

(D) $n + 3$

12. Look at the expression map below. Which label correctly identifies Part C?

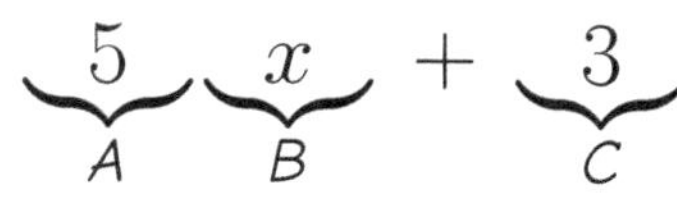

$$\underbrace{5}_{A} \ \underbrace{x}_{B} \ + \ \underbrace{3}_{C}$$

(A) A variable

(B) A coefficient

(C) A constant

(D) A factor

13. Evaluate $\dfrac{x^2}{4}$ when $x = 8$.

Your Answer

14. Simplify: $3(x + 4) + 2x$

(A) $5x + 4$

(B) $5x + 12$

(C) $6x + 12$

(D) $3x + 6$

15. Emma earns \$8 per hour babysitting. She also got a \$15 tip. Which expression gives her total earnings for h hours?

(A) $8 + 15h$

(B) $23h$

(C) $8h + 15$

(D) $8h - 15$

16. *Solve:* $x + 3.5 = 10$

(A) $x = 6.5$ (B) $x = 13.5$

(C) $x = 7.5$ (D) $x = 3.5$

17. Write an inequality: A suitcase must weigh less than 50 pounds. Let w = weight.

Your Answer:

18. Pick two values that ARE solutions to $x > -1$ and one value that is NOT.

Your Answer:

19. A car wash charges \$10 per car. Which variable goes on the x-axis when graphing?

(A) Total cost   (B) Number of cars

(C) Price per car   (D) Profit

20. A triangle has base $\frac{3}{4}$ ft and height $\frac{2}{3}$ ft. What is its area?

Your Answer

21. Which pair of points forms a vertical segment?

(A) $(3, 2)$ and $(7, 2)$ (B) $(5, -1)$ and $(5, 4)$

(C) $(1, 3)$ and $(4, 6)$ (D) $(0, 0)$ and $(3, 0)$

Find more at
ViewMath.com/AK-Grade6

22. A rectangle has vertices $(2, -1)$, $(2, 5)$, $(9, 5)$, and $(9, -1)$. What is the area?

23. A rectangular prism has length 8 cm, width 5 cm, and height 2 cm. Leo says the surface area is 80 cm^2. What did Leo calculate instead?

(A) The volume

(B) The perimeter

(C) Half the surface area

(D) The area of just one face

24. Which of the following is a statistical question?

(A) What year was the school built?

(B) What is the name of our principal?

(C) How many minutes does each student exercise per day?

(D) How many continents are there?

25. The mean of 4 numbers is 25. What is the sum of the 4 numbers?

(A) 25

(B) 50

(C) 75

(D) 100

26. Team A scores: $60, 62, 64, 66, 68$. Team B scores: $40, 50, 64, 78, 88$. Both have the same mean. Which team is more consistent?

(A) Team A, because its range is smaller.

(B) Team B, because its range is smaller.

(C) They are equally consistent because their means are equal.

(D) Neither is consistent.

Find more at
ViewMath.com/AK-Grade6

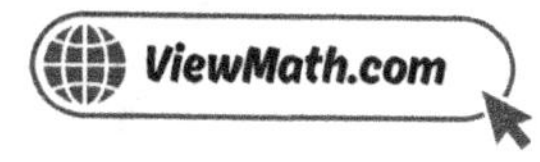

27. *A dot plot shows data clustered from 20 to 25 with no dots at 26, 27, 28, 29, and one dot at 30. What feature exists between 25 and 30?*

(A) *A cluster*

(B) *A peak*

(C) *A gap*

(D) *An interval*

28. *Data (in order):* $3, 5, 7, 9, 11, 13, 15.$ *What is Q1?*

(A) 3

(B) 5

(C) 7

(D) 9

29. *A data set has a range of 40 but an IQR of only 8. What does this tell you?*

(A) *All values are close together.*

(B) *Most values are clustered, but there are extreme values far from the center.*

(C) *The median equals the mean.*

(D) *The data is symmetric.*

30. *A frequency table shows that 5 students chose pizza, 8 chose tacos, 3 chose pasta, and 4 chose salad. How many students were surveyed in all?*

(A) 16

(B) 18

(C) 20

(D) 24

Find more at
ViewMath.com/AK-Grade6

 End of Practice Test 6 

Great job finishing the test!

 My Score

I got _____________ out of 30 questions right.

Check your answers in the Answer Key at the back of the book.

💡 Review any questions you missed. That's how we learn!

📊 Check Your Score Online!

Visit **ViewMath Academy** to enter your answers and see which topics you need to review. You can also explore lessons, take quizzes, track your scores, and save your progress!

viewmath.com/score/6.1.AK.21

Or go to viewmath.com/score and enter code: 6.1.AK.21

Practice Test 7

30 Questions

✏️ Before You Start ✏️

- ✔ **Read each question carefully** before choosing your answer.
- ✔ **Show your work** on scratch paper when you need to.
- ✔ **Skip hard questions** and come back to them later.
- ✔ **Check your answers** when you're done.
- ✔ **Take your time** — there's no rush!

⭐ You've Got This! ⭐

Do your best and show what you know!

1. A lemonade stand sells 7 cups of lemonade **for each** 2 cups of iced tea. If they sold 21 cups of lemonade, how many cups of iced tea did they sell?

Your Answer:

2. A garden hose fills a 150-gallon tank in 5 hours. What is the unit rate?

(A) 25 gallons per hour

(B) 30 gallons per hour

(C) 35 gallons per hour

(D) 750 gallons per hour

3. Complete the ratio table for the ratio 5 : 8.

5	8
10	?
?	32

Your Answer:

4. A class has 30 students. 70% passed a quiz. How many students passed?

(A) 7

(B) 21

(C) 23

(D) 70

5. Two families split a $350 dinner bill in a 3 : 4 ratio. How much does each family pay?

Your Answer:

Find more at
ViewMath.com/AK-Grade6

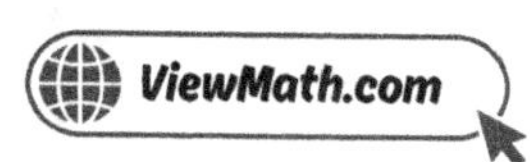

6. *What is $\frac{1}{2} \div \frac{1}{4}$?*

(A) $\frac{1}{8}$

(B) 8

(C) 2

(D) $\frac{1}{2}$

7. *Compute $9{,}072 \div 42$.*

Your Answer:

8. *Sam buys 3 folders that each cost $2.75. How much does he spend in all?*

(A) $8.75

(B) $8.25

(C) $6.25

(D) $5.75

9. *What is the opposite of 5?*

(A) 5

(B) -5

(C) 0

(D) $\frac{1}{5}$

10. *Where is the point $(-6, 0)$ located on the coordinate plane?*

(A) On the y-axis

(B) On the x-axis

(C) In Quadrant II

(D) In Quadrant III

11. *Write a word phrase for the expression $5x + 3$.*

Your Answer:

Find more at
ViewMath.com/AK-Grade6

ViewMath.com

12. Which expression has exactly 4 terms?

   A) $2x + 3y$

   B) $a + b + c$

   C) $5m - 2n + 7 + m$

   D) $4p$

13. The perimeter of a rectangle is $P = 2l + 2w$. What is P when $l = 9$ and $w = 4$?

   A) 13

   B) 22

   C) 26

   D) 36

14. Simplify: $5p + 3 + 2p + 4p - 1$

   A) $11p + 2$

   B) $7p + 2$

   C) $11p + 4$

   D) $12p$

15. A taxi charges \$3 plus \$2 per mile. Write an expression for the cost of a ride that is m miles long.

Your Answer

16. Solve: $5n = 45$

   A) $n = 9$

   B) $n = 40$

   C) $n = 50$

   D) $n = 225$

17. Which inequality represents "a number x is greater than 7"?

   A) $x < 7$

   B) $x > 7$

   C) $x \leq 7$

   D) $x = 7$

18. A student graphs $x > 2$ with a closed circle at 2 and shading to the right. What is the error?

(A) Should shade to the left

(B) Should use an open circle at 2

(C) Should use a circle at 0 instead

(D) There is no error

19. A phone battery loses 5% each hour. Which is the independent variable?

(A) Battery percentage

(B) Phone model

(C) Number of hours

(D) Screen brightness

20. A triangular flower bed has a base of 4.5 m and a height of 6 m. What is its area?

Your Answer:

21. A triangle on the coordinate plane has vertices $(0,0)$, $(8,0)$, and $(4,6)$. What is the length of the base along the x-axis?

(A) 4 units

(B) 6 units

(C) 8 units

(D) 10 units

22. A square has vertices $(-2,-2)$, $(4,-2)$, $(4,4)$, and $(-2,4)$. What is the area?

(A) 24 square units

(B) 36 square units

(C) 12 square units

(D) 16 square units

23. Find the surface area of a rectangular prism with length 7 m, width 5 m, and height 3 m.

Your Answer:

Find more at
ViewMath.com/AK-Grade6

24. A student surveyed classmates about how many books they read last month and made the dot plot below.

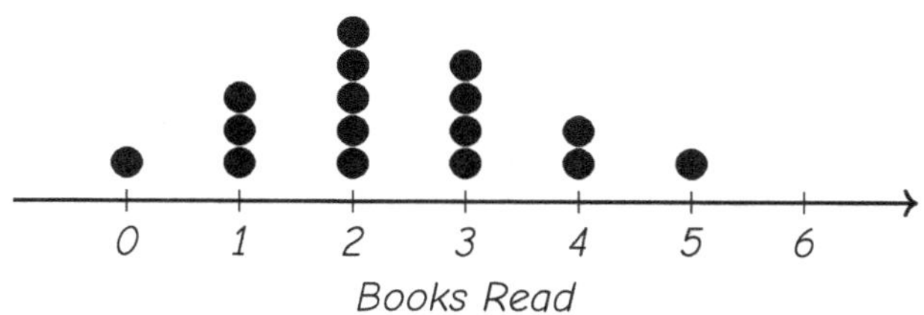

What does this dot plot confirm about the survey question "How many books did you read last month?"

(A) It is not statistical because only whole numbers appear.

(B) It is statistical because the answers vary from 0 to 5.

(C) It is not statistical because some students read the same number.

(D) It is statistical because there are exactly 16 dots.

25. Three villages reported moose counts: $15, 20, 25$. A fourth village reported 100 moose. Find the mean with and without the fourth village.

Your Answer

26. Which statement about MAD is true?

(A) MAD measures the middle value of the data.

(B) MAD tells you the average distance of each value from the mean.

(C) MAD is always larger than the range.

(D) MAD equals the range divided by 2.

27. *The dot plot below shows the scores on a 10-point quiz.*

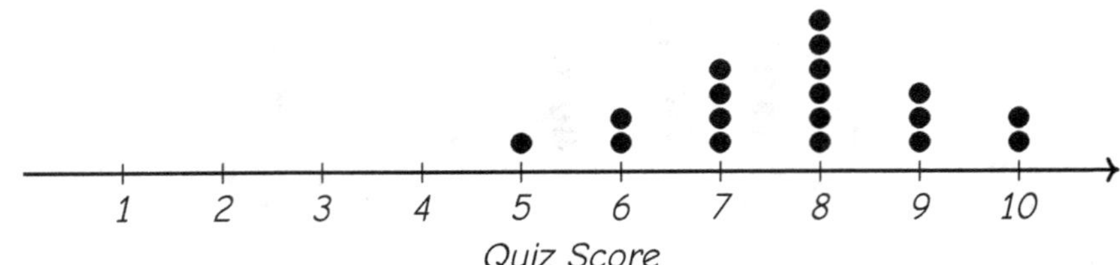

Find the total number of students, the median score, and describe the shape of the data.

Your Answer:

28. *Which five values make up the five-number summary?*

(A) Mean, median, mode, range, IQR

(B) Minimum, Q1, median, Q3, maximum

(C) Q1, Q2, Q3, Q4, Q5

(D) Mean, Q1, Q3, range, MAD

29. *Which situation would make the mean a poor measure of center?*

(A) A data set with all equal values

(B) A data set that is perfectly symmetric

(C) A data set with one value much larger than the rest

(D) A data set with an even number of values

30. *A double bar graph compares boys and girls who participated in a fun run. The bars show 24 boys and 30 girls. What fraction of all participants were boys?*

(A) $\dfrac{24}{30}$

(B) $\dfrac{4}{9}$

(C) $\dfrac{5}{9}$

(D) $\dfrac{1}{2}$

 End of Practice Test 7 

Great job finishing the test!

✅ My Score

I got __________ out of 30 questions right.

*Check your answers in the **Answer Key** at the back of the book.*

💡 Review any questions you missed. That's how we learn!

📊 Check Your Score Online!

Visit **ViewMath Academy** to enter your answers and see which topics you need to review. You can also explore lessons, take quizzes, track your scores, and save your progress!

viewmath.com/score/6.1.AK.22

Or go to viewmath.com/score and enter code: 6.1.AK.22

Practice Test 8

 30 Questions

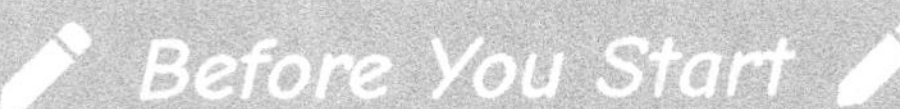

✏️ **Before You Start** ✏️

- ✔ **Read each question carefully** before choosing your answer.
- ✔ **Show your work** on scratch paper when you need to.
- ✔ **Skip hard questions** and come back to them later.
- ✔ **Check your answers** when you're done.
- ✔ **Take your time** — there's no rush!

 You've Got This! 

Do your best and show what you know!

1. The ratio of cats to dogs at a pet store is 3 : 4. Each part in the tape diagram represents 2 animals. How many dogs are there?

(A) 6

(B) 8

(C) 3

(D) 4

2. The graph shows the number of laps a swimmer completes over time.

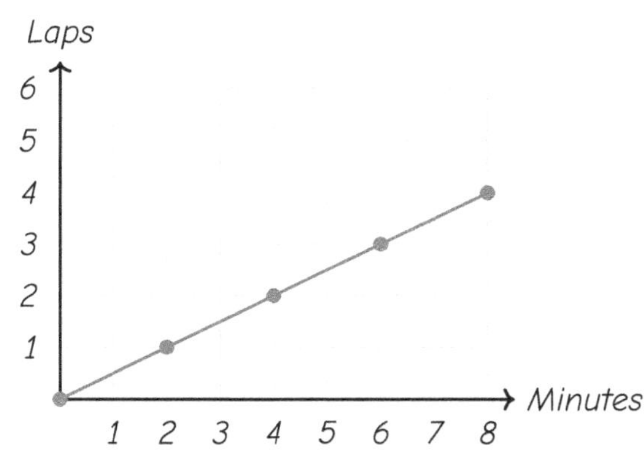

Part A: What is the swimmer's unit rate in laps per minute?

Part B: How many laps will the swimmer complete in 14 minutes?

Your Answer:

3. A smoothie recipe uses 2 cups of strawberries for every 3 cups of yogurt. How many cups of yogurt are needed for 10 cups of strawberries?

Your Answer:

4. Which of the following is 100% as a fraction?

(A) $\dfrac{1}{100}$

(B) $\dfrac{100}{10}$

(C) $\dfrac{100}{100}$

(D) $\dfrac{10}{100}$

 Find more at
ViewMath.com/AK-Grade6

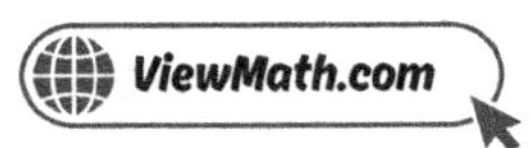

5. A printing company prints 250 flyers in 10 minutes. How long will it take to print 1,000 flyers?

Your Answer:

6. A student solved the problem below using "Keep, Change, Flip." Look at the student's work.

Problem: $\dfrac{4}{5} \div \dfrac{2}{3}$

Step 1: Keep $\dfrac{4}{5}$ Change $\div$ to $\times$

Step 2: Flip $\dfrac{4}{5}$ to get $\dfrac{5}{4}$

Step 3: $\dfrac{5}{4} \times \dfrac{2}{3} = \dfrac{10}{12} = \dfrac{5}{6}$

What error did the student make?

(A) The student forgot to change $\div$ to $\times$

(B) The student made a multiplication error in Step 3

(C) The student forgot to simplify

(D) The student flipped the wrong fraction

7. Which multiplication equation can be used to check that $4{,}752 \div 12 = 396$?

(A) $396 \times 12 = 4{,}752$

(B) $396 + 12 = 408$

(C) $4{,}752 \times 12 = 57{,}024$

(D) $396 \div 12 = 33$

8. What is 0.12×0.5?

(A) 0.6

(B) 0.06

(C) 0.006

(D) 6.0

Find more at
ViewMath.com/AK-Grade6

ViewMath.com

9. *What is the opposite of 0?*

(A) 1

(B) -1

(C) 0

(D) *There is no opposite of 0*

10. **Part A:** *In which quadrant is the point $(-8, 1)$ located?*

Part B: *If $(-8, 1)$ is reflected across the y-axis, what are the coordinates of its image and in which quadrant does it lie?*

Part C: *If the original point $(-8, 1)$ is instead reflected across the x-axis, what are the coordinates of its image and in which quadrant does it lie?*

Your Answer

11. *The table below shows the relationship between a word phrase and an expression. Which expression belongs in the blank?*

Word Phrase	Expression
A number plus 4	$n + 4$
3 times a number	$3n$
8 less than a number	?
A number divided by 2	$n \div 2$

(A) $8 - n$

(B) $n - 8$

(C) $8n$

(D) $n + 8$

12. *What is the coefficient of d in the expression $15 - d + 3$?*

(A) 0

(B) 1

(C) -1

(D) 15

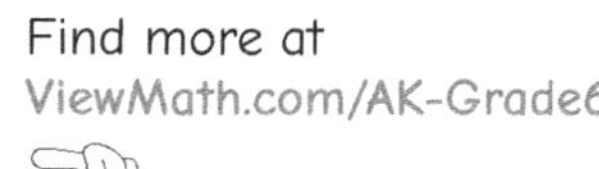

Find more at
ViewMath.com/AK-Grade6

13. A parking garage charges \$5 plus \$3 per hour. How much does it cost to park for 6 hours? Use the expression $5 + 3h$.

Your Answer

14. The rectangle below has the dimensions shown. Write a simplified expression for its perimeter.

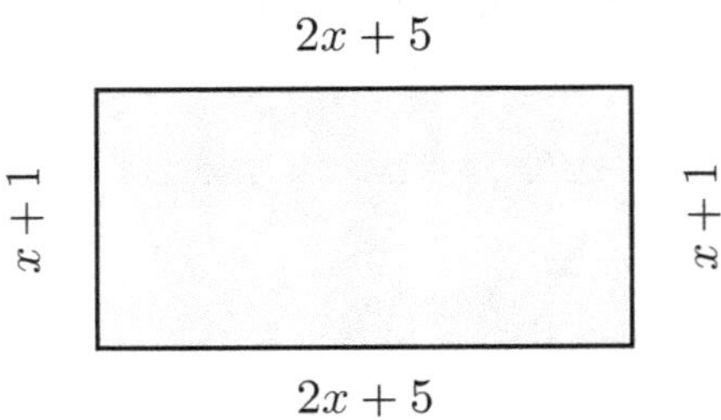

Your Answer

15. Marcus has n baseball cards. He buys 12 more and then gives 5 to his friend. Write an expression for how many cards he has now.

Your Answer

16. Solve: $15a = 105$

Your Answer

17. Is $n = 4$ a solution to $n > 4$? Is $n = 4$ a solution to $n \geq 4$? Explain both.

Your Answer

Find more at
ViewMath.com/AK-Grade6

18. *Which of these numbers is a solution to* $x \geq -3$*?*

(A) -4

(B) -3.5

(C) -3

(D) -100

19. *A bathtub drains at 3 gallons per minute. It starts with 45 gallons. Write an equation for the water w remaining after t minutes. When is the tub empty?*

Your Answer

20. *Look at the two triangles on the grid. Which triangle has a greater area?*

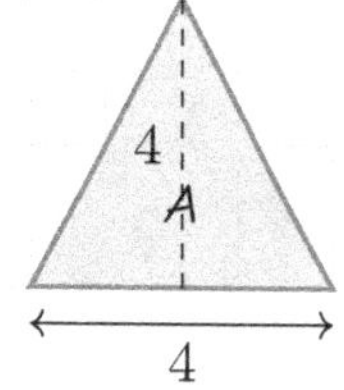

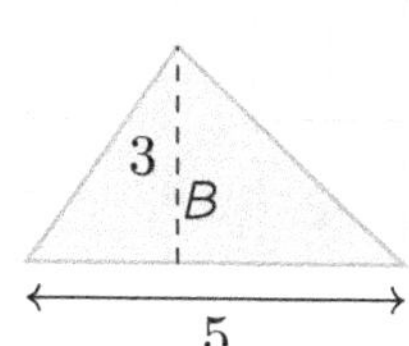

(A) Triangle A

(B) Triangle B

(C) They have the same area.

(D) Not enough information to tell.

21. *Three vertices of a rectangle are* $(2, 3)$*,* $(2, -5)$*, and* $(9, 3)$*. What is the fourth vertex?*

Your Answer

22. *A right triangle has vertices* $(0, 0)$*,* $(8, 0)$*, and* $(0, 6)$*. What is the area?*

(A) 48 *square units*

(B) 24 *square units*

(C) 14 *square units*

(D) 28 *square units*

23. What is a net?

(A) A 3D shape made of cubes

(B) A flat pattern that folds into a 3D shape

(C) The volume of a rectangular prism

(D) A grid used to measure area

24. Write a statistical question you could ask your classmates about food.

Your Answer:

25. A student measured snowfall (inches) over 4 months: $10, 14, 12, 16$. What score in a 5th month would make the mean exactly 13?

Your Answer:

26. Which measure of spread is most affected by outliers?

(A) IQR

(B) MAD

(C) Range

(D) Median

27. A dot plot of daily steps shows: 3000 (1 dot), 4000 (2 dots), 5000 (6 dots), 6000 (4 dots), 7000 (2 dots). What is the most common number of steps?

Your Answer:

28. Two box plots have the same median. Class A has IQR $= 8$ and Class B has IQR $= 24$. What can you conclude?

(A) Class A and Class B performed equally.

(B) Class A's scores are more spread out.

(C) Class B's middle 50% of scores are more spread out.

(D) The range of Class A is larger.

Find more at
ViewMath.com/AK-Grade6

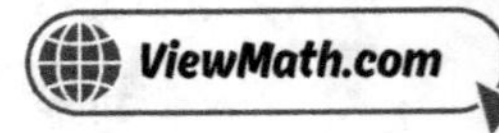

29. A student says "The data set with the larger range is always more spread out than the one with the smaller range." Is this always true? Explain.

Your Answer

30. In a frequency table, the colors chosen by students are: red (12), blue (7), green (7), yellow (4). How many more students chose red than yellow?

(A) 4

(B) 5

(C) 7

(D) 8

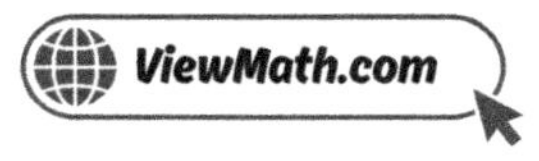

 # ★ End of Practice Test 8 ★ 

Great job finishing the test!

 My Score

I got _____________ out of 30 questions right.

Check your answers in the **Answer Key** at the back of the book.

Review any questions you missed. That's how we learn!

📊 Check Your Score Online!

Visit **ViewMath Academy** to enter your answers and see which topics you need to review. You can also explore lessons, take quizzes, track your scores, and save your progress!

viewmath.com/score/6.1.AK.23

Or go to viewmath.com/score and enter code: 6.1.AK.23

9

Practice Test 9

☑ 30 Questions

✏️ Before You Start ✏️

- ✓ **Read each question carefully** before choosing your answer.
- ✓ **Show your work** on scratch paper when you need to.
- ✓ **Skip hard questions** and come back to them later.
- ✓ **Check your answers** when you're done.
- ✓ **Take your time** — there's no rush!

⭐ You've Got This! ⭐

Do your best and show what you know!

1. The ratio of red to blue to green beads is $1 : 3 : 2$. There are 18 beads total. How many blue beads are there?

(A) 3

(B) 6

(C) 9

(D) 12

2. A printer prints 120 pages in 4 minutes. What is the unit rate?

(A) 4 pages per minute

(B) 30 pages per minute

(C) 120 pages per minute

(D) 480 pages per minute

3. Look at this ratio table. What is the missing value?

Apples	Oranges
3	5
6	?

(A) 8

(B) 10

(C) 15

(D) 11

4. A test has 100 questions. Emma got 88 correct. What percent did she get right?

(A) 12%

(B) 8.8%

(C) 88%

(D) 78%

5. A batch of trail mix uses 4 cups of granola for every 3 cups of raisins. If you have 12 cups of raisins, how many cups of granola do you need?

(A) 9

(B) 12

(C) 16

(D) 15

Find more at
ViewMath.com/AK-Grade6

6. A ribbon is $\frac{3}{4}$ yard long. You cut pieces that are each $\frac{1}{8}$ yard. How many pieces can you cut?

(A) 6

(B) $\frac{3}{32}$

(C) 4

(D) 8

7. What is $2{,}688 \div 16$?

(A) 158

(B) 168

(C) 178

(D) 1,068

8. The table shows the distances a cyclist rode each day.

Day	Distance (km)
Monday	8.76
Tuesday	12.30
Wednesday	6.48

Part A: What is the total distance the cyclist rode over the three days?

Part B: If the cyclist wants to divide the total distance equally over the 3 days, how many kilometers per day is that?

Your Answer

9. A number plus its opposite always equals which value?

(A) 1

(B) The number itself

(C) -1

(D) 0

Find more at
ViewMath.com/AK-Grade6

10. *A triangle is drawn on the coordinate plane below with vertices at A, B, and C.*

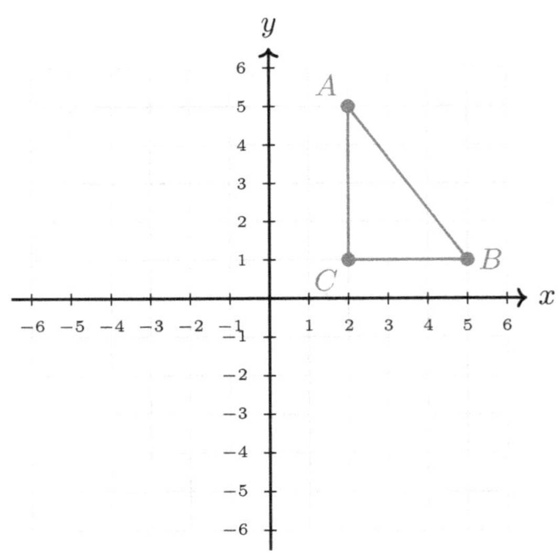

Part A: *Write the coordinates of each vertex: A, B, and C.*

Part B: *If the triangle is reflected across the y-axis, write the coordinates of the new vertices A', B', and C'.*

Part C: *In which quadrant will the reflected triangle be located?*

Your Answer:

---

11. *Which expression represents "the sum of 9 and the product of 4 and t"?*

(A) $9 + 4 + t$

(B) $9 + 4t$

(C) $9(4 + t)$

(D) $9 \times 4t$

---

12. *How many terms are in the expression $4x + 9 - 2y$?*

(A) 2

(B) 3

(C) 4

(D) 5

13. Evaluate $6x - x^2$ when $x = 4$

    (A) 8   (B) 20

    (C) -8 (D) 40

14. Simplify: $7n + 4n$

    (A) $11n$ (B) $28n$

    (C) $11n^2$ (D) $74n$

15. In $P = 4s$, can s be any positive number, or is it one specific number?

    (A) Only $s = 1$ (B) Only whole numbers

    (C) Any positive number — the formula works for   (D) Only $s = 4$
all squares

16. Solve: $\dfrac{t}{6} = 5$

    (A) $t = 11$ (B) $t = 1$

    (C) $t = 30$ (D) $t = 56$

17. Is $x = 5$ a solution to $x > 5$?

    (A) Yes, because 5 is equal to 5   (B) Yes, because 5 is positive

    (C) No, because 5 is not greater than 5   (D) No, because 5 is greater than 5

18. Which inequality matches a graph with an open circle at -2 and shading to the left?

   (A) $x > -2$ (B) $x < -2$

   (C) $x \geq -2$ (D) $x \leq -2$

19. In $c = 7p$, where c is total cost and p is the number of pizzas, what is the cost of 6 pizzas?

   (A) \$13   (B) \$36

   (C) \$42   (D) \$67

20. A triangle has base $\frac{1}{2}$ ft and height $\frac{1}{4}$ ft. What is the area?

   (A) $\frac{1}{8}$ ft^2 (B) $\frac{3}{4}$ ft^2

   (C) $\frac{1}{16}$ ft^2 (D) $\frac{1}{4}$ ft^2

21. A square has one vertex at $(-2, -2)$ and side length 5. The sides are horizontal and vertical. Which could be the opposite vertex?

   (A) $(3, 3)$ (B) $(3, -7)$

   (C) $(5, 5)$ (D) $(2, 2)$

22. A rectangle has vertices $(0, 0)$, $(6, 0)$, $(6, 4)$, and $(0, 4)$. What is the area?

   (A) 10 square units   (B) 20 square units

   (C) 24 square units   (D) 12 square units

Find more at
ViewMath.com/AK-Grade6

ViewMath.com

23. A cube has a surface area of 96 cm². What is the edge length?

(A) 3 cm

(B) 4 cm

(C) 6 cm

(D) 8 cm

24. Which question is **not** statistical?

(A) How many siblings do students in our school have?

(B) How far do students travel to get to school?

(C) What is 12×9?

(D) How many pets do families in our neighborhood own?

25. The dot plot below shows the number of fish caught by 10 fishers on one day.

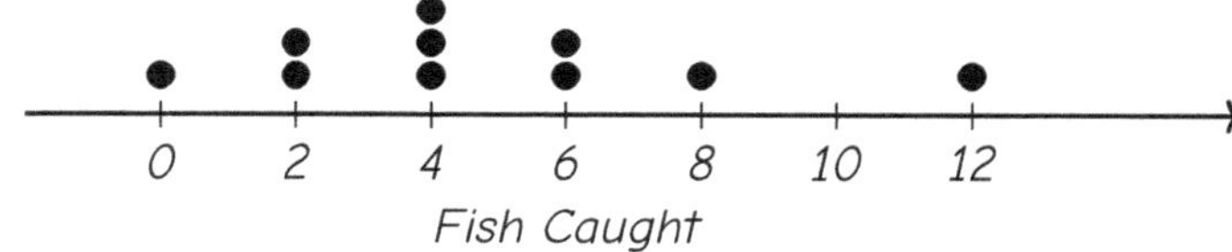

What is the median number of fish caught?

(A) 2

(B) 4

(C) 5

(D) 6

26. If every value in a data set is increased by 5, what happens to the range?

(A) It increases by 5.

(B) It decreases by 5.

(C) It stays the same.

(D) It doubles.

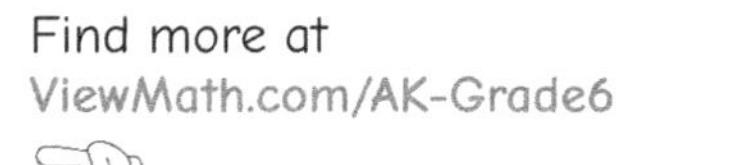
Find more at
ViewMath.com/AK-Grade6

27. Look at the histogram below showing the ages of participants in a race.

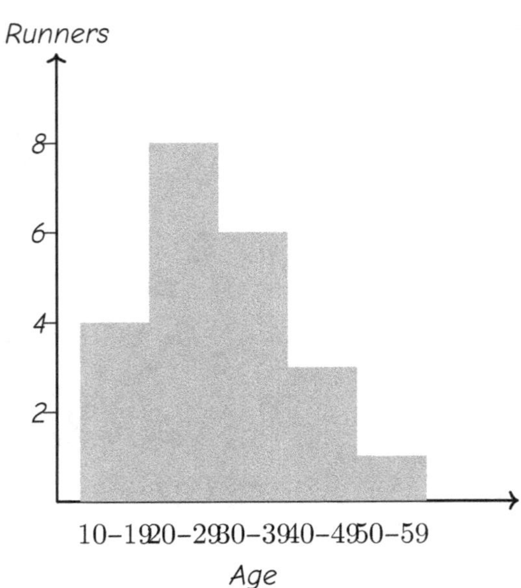

What fraction of runners are in the 20–29 age group?

(A) $\frac{4}{22}$ (B) $\frac{8}{22}$

(C) $\frac{8}{14}$ (D) $\frac{6}{22}$

28. Data (in order): $2, 4, 6, 8, 10, 12$. What is the median?

(A) 6   (B) 7

(C) 8   (D) 10

29. Data set X: $10, 20, 30, 40, 50, 60, 70$. Data set Y: $35, 37, 38, 40, 42, 43, 45$. Both have median 40. Which data set has a smaller IQR?

(A) Data set X   (B) Data set Y

(C) They are the same.   (D) Cannot be determined.

Find more at
ViewMath.com/AK-Grade6

30. *A double bar graph shows the number of apples and oranges sold each day. On Monday, 15 apples and 20 oranges were sold. On Tuesday, 25 apples and 10 oranges were sold. On which day were more total fruits sold, and how many more?*

Your Answer:

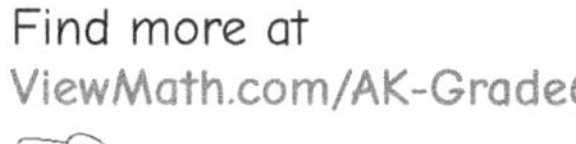

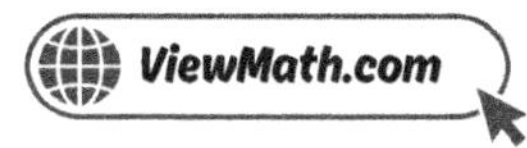

 End of Practice Test 9 

Great job finishing the test!

 My Score

I got _____________ out of 30 questions right.

*Check your answers in the **Answer Key** at the back of the book.*

💡 *Review any questions you missed. That's how we learn!*

📊 Check Your Score Online!

Visit **ViewMath Academy** to enter your answers and see which topics you need to review. You can also explore lessons, take quizzes, track your scores, and save your progress!

viewmath.com/score/6.1.AK.24

Or go to viewmath.com/score and enter code: 6.1.AK.24

10

Practice Test 10

📋 30 Questions

✏️ Before You Start ✏️

- ✓ **Read each question carefully** before choosing your answer.
- ✓ **Show your work** on scratch paper when you need to.
- ✓ **Skip hard questions** and come back to them later.
- ✓ **Check your answers** when you're done.
- ✓ **Take your time** — there's no rush!

⭐ You've Got This! ⭐

Do your best and show what you know!

1. The tape diagram below shows the ratio of apple juice to orange juice in a drink.

Apple: ▢▢▢

Orange: ▢▢▢▢▢

If each part equals 4 ounces, how many total ounces of drink are there?

(A) 12

(B) 20

(C) 32

(D) 8

2. The table below shows the prices at two stores for packs of pencils.

	Number of Pencils	Price
Store A	10	$4.50
Store B	8	$3.20

Which store has the lower unit price per pencil?

(A) Store A at $0.45 per pencil

(B) Store B at $0.40 per pencil

(C) Store A at $0.40 per pencil

(D) They have the same unit price.

3. Which pair of ratios are equivalent?

(A) 2 : 3 and 4 : 9

(B) 2 : 3 and 6 : 9

(C) 2 : 3 and 8 : 9

(D) 2 : 3 and 3 : 2

4. Order from least to greatest: $\frac{1}{3}$, 30%, 0.35.

(A) 30%, $\frac{1}{3}$, 0.35

(B) $\frac{1}{3}$, 30%, 0.35

(C) 0.35, 30%, $\frac{1}{3}$

(D) 30%, 0.35, $\frac{1}{3}$

Find more at
ViewMath.com/AK-Grade6

ViewMath.com

5. A fence uses posts in a ratio of 1 post for every 6 feet. How many posts are needed for a 42-foot fence?

(A) 6

(B) 7

(C) 8

(D) 36

6. Bar A and Bar B below represent the same whole. Bar A has $\frac{2}{3}$ shaded. Bar B has $\frac{1}{6}$ shaded.

Bar A

$\frac{1}{3}$	$\frac{1}{3}$	$\frac{1}{3}$

Bar B

$\frac{1}{6}$	$\frac{1}{6}$	$\frac{1}{6}$	$\frac{1}{6}$	$\frac{1}{6}$	$\frac{1}{6}$

How many times does the shaded part of Bar B fit into the shaded part of Bar A? Write a division equation and solve.

Your Answer:

7. Compute $3{,}780 \div 15$.

Your Answer:

8. Compute $8.64 \div 0.4$.

Your Answer:

9. What is $|0|$?

(A) There is no answer

(B) -1

(C) 1

(D) 0

10. *In which quadrant is the point $(2, -5)$ located?*

(A) *Quadrant I*

(B) *Quadrant II*

(C) *Quadrant III*

(D) *Quadrant IV*

11. *Look at the balance model below. Each triangle represents the same unknown number x, and each small square represents 1. Write an expression for the total value shown on the balance.*

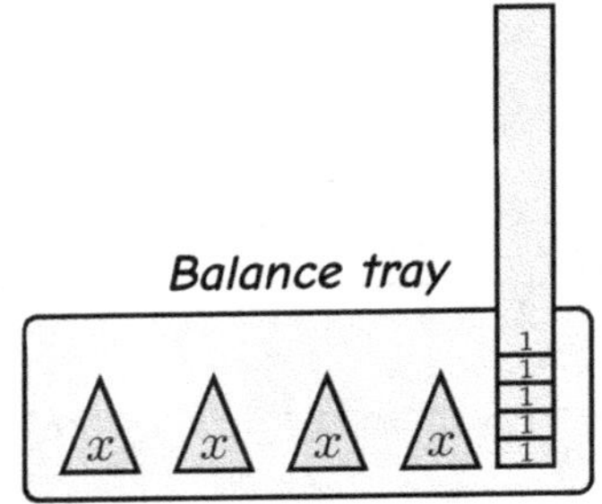

Your Answer

12. *In the term $\dfrac{n}{5}$, what are the factors?*

(A) *n and 5*

(B) *n and $\dfrac{1}{5}$*

(C) *1 and n*

(D) *5 and $\dfrac{1}{n}$*

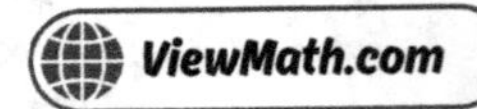

13. The table below shows the cost of renting a bike using the expression $C = 8 + 3h$, where h is the number of hours. Fill in the missing values A and B.

Hours (h)	Cost (C)
1	$11
2	A
3	$17
5	B

Your Answer:

14. Which expression is equivalent to $7(k - 3) + 5$?

A) $7k - 16$

B) $7k + 2$

C) $7k - 26$

D) $7k - 21$

15. You have 100 pages to read and read r pages each day. Which expression tells how many pages are left after 4 days?

A) $100 + 4r$

B) $100 - 4r$

C) $4r - 100$

D) $100r - 4$

16. Solve: $m + 17 = 30$

Your Answer:

17. Which inequality represents "you need more than \$25 to buy the video game"?

(A) $d \leq 25$ (B) $d < 25$

(C) $d > 25$ (D) $d \geq 25$

18. Look at the number line below and write the inequality it represents.

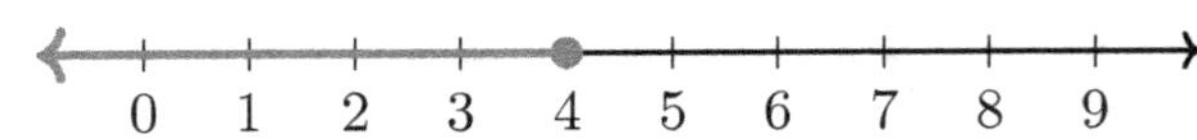

Your Answer:

19. Which equation represents this relationship: "The number of legs L is 4 times the number of dogs d"?

(A) $d = 4L$ (B) $L = d + 4$

(C) $L = 4d$ (D) $L = d \div 4$

20. A triangle has an area of 24 ft^2 and a base of 8 ft. What is the height?

(A) 3 ft   (B) 6 ft

(C) 16 ft   (D) 4 ft

21. What is the distance between the points $(2, 5)$ and $(2, -3)$?

(A) 2 units   (B) 5 units

(C) 8 units   (D) 3 units

22. An L-shaped figure has vertices $(0,0)$, $(8,0)$, $(8,3)$, $(4,3)$, $(4,7)$, and $(0,7)$. A student splits it into two rectangles. Which pair of rectangles correctly covers the figure?

(A) 8×7 and 4×3

(B) 8×3 and 4×4

(C) 4×7 and 4×3

(D) 8×3 and 4×7

23. A triangular prism has two equilateral triangle bases with side 6 cm and height 5.2 cm. The prism is 12 cm long. What is the total surface area?

Your Answer

24. Look at the flowchart below. Which path correctly identifies a statistical question?

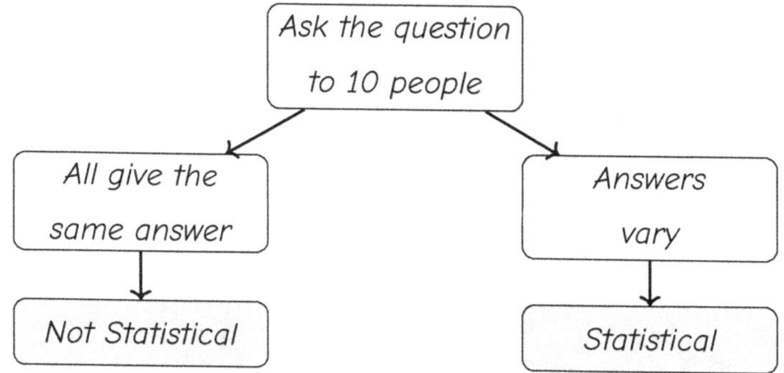

Using the flowchart, which question would end at "Statistical"?

(A) How many wheels does a bicycle have?

(B) What is the sum of $10 + 5$?

(C) How many minutes do you spend eating lunch?

(D) How many days are in one week?

25. Caribou spotted per day on a 7-day survey: $8, 12, 10, 14, 11, 9, 13$. Find the median.

(A) 10

(B) 11

(C) 12

(D) 14

Find more at
ViewMath.com/AK-Grade6

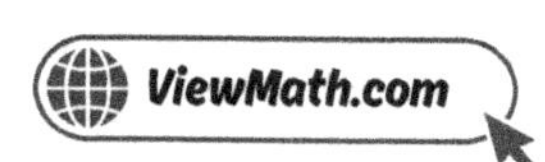

26. Two data sets have the same mean of 50. Data set A: MAD = 2. Data set B: MAD = 10. Which data set has more variability?

Your Answer

27. How is a histogram different from a bar graph?

(A) A histogram uses circles instead of bars.

(B) A histogram shows data grouped into intervals with no gaps between bars.

(C) A histogram can only show 10 data values.

(D) A histogram always has equal-height bars.

28. Look at the box plot below.

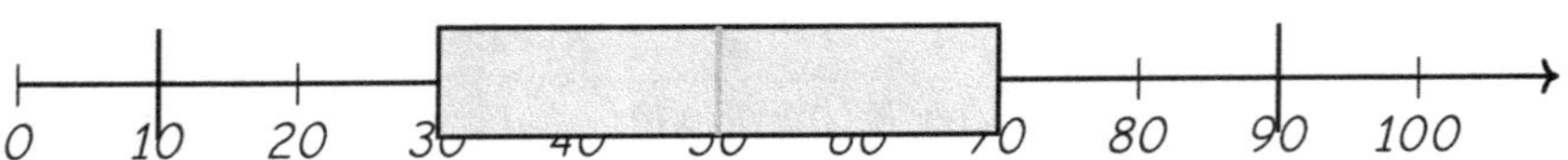

What is the IQR of this data?

(A) 20

(B) 40

(C) 50

(D) 80

29. You are writing a report about two data sets. Data set 1 has a symmetric shape; data set 2 is skewed. Which center measure should you use for each?

(A) Mean for both

(B) Median for both

(C) Mean for symmetric; median for skewed

(D) Mode for both

Find more at
ViewMath.com/AK-Grade6

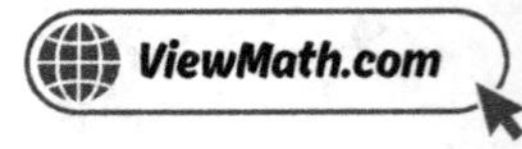

30. A frequency table shows test scores. The title of the table is "Math Quiz Scores for Period 3." What does the title tell you?

(A) The scores are from a science test

(B) The scores are from every class in the school

(C) The scores are math quiz results from one specific class period

(D) The table shows grades for the whole year

 End of Practice Test 10 

Great job finishing the test!

☑ My Score

I got ___________ out of 30 questions right.

Check your answers in the Answer Key at the back of the book.

💡 *Review any questions you missed. That's how we learn!*

📊 Check Your Score Online!

Visit **ViewMath Academy** to enter your answers and see which topics you need to review. You can also explore lessons, take quizzes, track your scores, and save your progress!

viewmath.com/score/6.1.AK.25

Or go to viewmath.com/score and enter code: 6.1.AK.25

Answer Key & Explanations

Answer Key

First try each test on your own, then check your work here.

✅ Practice Test 1 — Answer Key

1 5   **2** C   **3** C   **4** D   **5** C   **6** 4   **7** C   **8** A   **9** 15

10 $(-7, -3)$ **11** B   **12** $A = 4, B = (x + 3)$ **13** 22   **14** B

15 Answers vary. Example: You buy n notebooks at \$5 each and pay \$3 for shipping.   **16** C

17 Answers vary. Example: A classroom has at most 15 computers.

18 $x = 3$ IS a solution to $x \leq 3$ because $3 \leq 3$ is true. $x = 3$ is NOT a solution to $x < 3$ because $3 < 3$ is false.

19 B   **20** C   **21** D   **22** B   **23** B   **24** B   **25** B   **26** C   **27** D

28 Range $= 55$, IQR $= 30$ **29** B   **30** positive trend

💡 Time to Learn! 💡

*Review the explanations below, **especially for the questions you missed**.*

Understanding why each answer is correct builds stronger problem-solving skills.

Tip: *Circle any questions you got wrong, then read their explanation carefully.*

Find more at
ViewMath.com/AK-Grade6

🌐 **ViewMath.com**

Practice Test 1 — Detailed Explanations

1. Each part $= 20 \div 4 = 5$. Vegetables $= 1 \times 5 = 5$.

2. From the graph, the truck travels 50 miles in 2 hours. Unit rate: $50 \div 2 = 25$ miles per hour.

3. From the double number line, 9 scoops aligns with 15 cups of water.

4. Divide by 100: $72 \div 100 = 0.72$.

5. Scale: 2 cm $= 15$ km, so 1 cm $= 7.5$ km. For 5 cm: $5 \times 7.5 = 37.5$ km.

6. $2\frac{1}{2} = \frac{5}{2}$. Then $\frac{5}{2} \div \frac{5}{8} = \frac{5}{2} \times \frac{8}{5} = \frac{40}{10} = 4$ shelves.

7. Total cans: $384 + 528 + 456 + 432 = 1{,}800$. Divide: $1{,}800 \div 12 = 150$ cans per classroom.

8. Multiply: 5.4×6. As whole numbers: $54 \times 6 = 324$. One decimal place: 32.4 liters.

9. $|-9| = 9$ and $|-6| = 6$. Adding these gives $9 + 6 = 15$. Find each absolute value first, then add.

10. Reflecting across the x-axis changes the sign of the y-coordinate: $(x, y) \to (x, -y)$. So $(-7, 3)$ becomes $(-7, -3)$.

11. Sandwiches cost $6s$ and drinks cost $2d$. Total: $6s + 2d$.

12. $4(x + 3) = 4 \times (x + 3)$. The two factors are 4 and $(x + 3)$.

Find more at
ViewMath.com/AK-Grade6

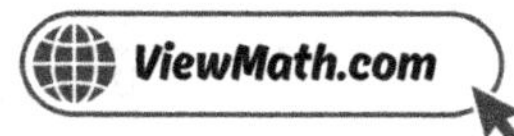

13 $3(8 + 2) - 8 = 3(10) - 8 = 30 - 8 = 22$.

14 Distribute: $3 \times x + 3 \times 5 = 3x + 15$.

15 $5n$ represents a cost per item and 3 represents a fixed fee.

16 Multiply both sides by 4: $m = 7 \times 4 = 28$.

17 Any situation where a quantity is 15 or fewer works.

18 $\leq$ includes the boundary value; $<$ does not.

19 In B, y is always 7 no matter what x is. The value of y does not change with x.

20 One sail: $\frac{1}{2} \times 3 \times 7 = 10.5 \ m^2$. Two sails: $10.5 \times 2 = 21 \ m^2$.

21 $|4 - (-5)| = |9| = 9$ units.

22 Base $= |2 - (-4)| = 6$. Height $= |5 - 0| = 5$. Area $= \frac{1}{2} \times 6 \times 5 = 15$ square units.

23 $SA = 2(10)(3) + 2(10)(7) + 2(3)(7) = 60 + 140 + 42 = 242 \ m^2$.

24 The answers range from 0 to 4 and vary from person to person, which confirms it is a statistical question.

25 Mean $= (15 + 18 + 20 + 22 + 25) \div 5 = 100 \div 5 = 20$ miles.

26 MAD $= (4 + 2 + 0 + 2 + 4) \div 5 = 12 \div 5 = 2.4$.

Find more at
ViewMath.com/AK-Grade6

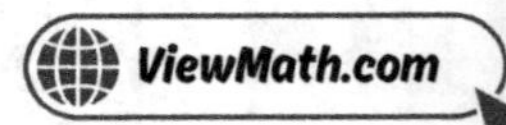

27 The value at 12 is far from the cluster (6–8) with a gap in between, making it an outlier.

28 Range $= 80 - 25 = 55$. IQR $= 65 - 35 = 30$.

29 School B's IQR (45) is larger than School A's (30), indicating more variability in the middle 50% of scores.

30 Since the temperature increases each month from January to May, the line goes upward, which is a positive trend.

✅ Practice Test 2 — Answer Key

1 C   2 Part A: Lily earns \$12 per hour, Noah earns \$13 per hour.   Part B: Noah earns \$1 more per hour.

3 Part A: 2 : 1;   Part B: 4 cups of sugar   4 C   5 B   6 D   7 A   8 \$16.43

9 B

10 The point $(9, 0)$ lies on the x-axis because its y-coordinate is 0. Points on the axes are not inside any quadrant.

11 C   12 5 and $(y + 8)$ 13 30 square cm   14 $6a + 10$ 15 A   16 B   17 C

18 C   19 Independent: h (hours). Dependent: d (distance).   20 D   21 A   22 64 square units

23 C   24 No   25 B   26 Range $= 7$, IQR $= 2$ 27 C   28 D   29 A   30 B

Find more at
ViewMath.com/AK-Grade6

> 💡 **Time to Learn!** 💡
>
> Review the explanations below, **especially for the questions you missed**.
>
> Understanding why each answer is correct builds stronger problem-solving skills.
>
> **Tip:** Circle any questions you got wrong, then read their explanation carefully.

📖 Practice Test 2 — Detailed Explanations

1 Girls have 6 parts $= 24$, so each part $= 24 \div 6 = 4$. Boys have 4 parts $= 4 \times 4 = 16$.

2 Lily: $\$60 \div 5 = \$12/hr$. Noah: $\$52 \div 4 = \$13/hr$. Noah earns $\$13 - \$12 = \$1$ more per hour.

3 Part A: From the graph, $(2, 1)$ shows 2 cups of flour for every 1 cup of sugar. Part B: Following the pattern, $(8, 4)$, so 4 cups of sugar.

4 $50\% = \dfrac{50}{100} = \dfrac{1}{2}$.

5 $360 \div 12 = 30$ minutes.

6 $\dfrac{7}{8} \div \dfrac{1}{4} = \dfrac{7}{8} \times \dfrac{4}{1} = \dfrac{28}{8} = \dfrac{7}{2} = 3\dfrac{1}{2}$ laps.

7 $7 \div 3 = 2$ R1; bring down $5 \rightarrow 15 \div 3 = 5$; bring down $6 \rightarrow 6 \div 3 = 2$. Answer: 252. Check: $252 \times 3 = 756$.

8 $12.95 + 3.48 = 16.43$. Hundredths: $5 + 8 = 13$, write 3, carry 1. Tenths: $9 + 4 + 1 = 14$, write 4, carry 1. Ones: $2 + 3 + 1 = 6$. Tens: 1.

Find more at
ViewMath.com/AK-Grade6

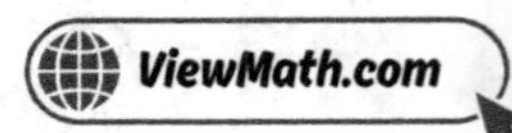

9. Absolute value tells you how far a number is from zero. -12 is 12 units from zero, so $|-12| = 12$. A common mistake is to think $|-12| = -12$, but distance is never negative.

10. The four quadrants are the regions between the axes. A point is on an axis when $x = 0$ (on the y-axis) or $y = 0$ (on the x-axis). Since $(9, 0)$ has $y = 0$, it sits on the x-axis itself, which is the boundary between Quadrant I and Quadrant IV—not in either one.

11. Equally distributing c cupcakes into 6 boxes means dividing: $c \div 6$.

12. $5(y + 8) = 5 \times (y + 8)$. The factors are 5 and $(y + 8)$.

13. $A = \frac{1}{2} \times 10 \times 6 = \frac{1}{2} \times 60 = 30$ square cm.

14. $3a + 5a - 2a = 6a$ and $7 + 3 = 10$. Result: $6a + 10$.

15. Tickets: $9p$. One bucket of popcorn: 6. Total: $9p + 6$.

16. Subtract 15: $k = 32 - 15 = 17$.

17. "Fewer than 20" means less than 20: $y < 20$.

18. Open circle at 0 means 0 is not included. Shading right means greater than: $x > 0$.

19. You choose the hours (independent). Distance depends on hours: $d = 6h$.

20. $A = \frac{1}{2} \times 14 \times 8 = 56$ m^2.

21. From $(0, 0)$ with length 10 along the x-axis and width 4 along the y-axis, the opposite vertex is $(10, 4)$.

Find more at
ViewMath.com/AK-Grade6

22 $Side = |5 - (-3)| = 8.$ $Area = 8 \times 8 = 64$ square units.

23 A rectangular prism has 6 faces: 3 pairs of opposite rectangles.

24 "How many inches in a foot?" has a numerical answer (12) but is not statistical because the answer does not vary. A statistical question must produce data that varies.

25 $Mean = (-10 + 10 + 60 + 55) \div 4 = 115 \div 4 = 28.75°F.$

26 Data: $3, 4, 4, 5, 5, 5, 6, 6, 6, 6, 7, 7, 8, 10.$ 14 values. $Range = 10 - 3 = 7.$ $Median = (6 + 6) \div 2 = 6.$ Lower half: $3, 4, 4, 5, 5, 5, 6,$ $Q1 = 5.$ Upper half: $6, 6, 7, 7, 8, 10...$ wait, let me recount. 14 values: lower 7: $3, 4, 4, 5, 5, 5, 6$ ⊠ $Q1 = 5.$ Upper 7: $6, 6, 6, 7, 7, 8, 10$ ⊠ $Q3 = 7.$ $IQR = 7 - 5 = 2.$

27 A dot plot places one dot for each data value, so you can see every individual value.

28 Different data sets can have the same five-number summary. The five-number summary only captures five key values, not every data point.

29 Class 1's scores cluster tightly (75–85), giving a small IQR despite the outlier at 40. Class 2's even spread gives a large IQR. Class 1 is more consistent.

30 Boys in Class A: bar reaches 6 units = 30 books. Girls in Class A: bar reaches 4 units = 20 books. Difference: $30 - 20 = 10.$

✓ Practice Test 3 — Answer Key

1	C	2	B	3	B	4	C	5	B	6	B	7	C	8	B	9	C	10	B

11	D	12	1	13	C	14	$8y + 24$	15	B	16	A	17	D	18	B	19	B

Find more at
ViewMath.com/AK-Grade6

 **20** C    **21** Б    **22** 28 square units    **23** A    **24** B    **25** B    **26** 6    **27** A    **28** B

 **29** B    **30** B

💡 Time to Learn! 💡

*Review the explanations below, **especially for the questions you missed**.*

Understanding why each answer is correct builds stronger problem-solving skills.

***Tip:** Circle any questions you got wrong, then read their explanation carefully.*

📖 Practice Test 3 — Detailed Explanations

1. The question asks flour to eggs. Flour $= 3$, eggs $= 2$. The ratio is $3 : 2$.

2. $\$9 \div 12 = \$0.75$ per muffin.

3. $7 \times 3 = 21$ bananas, so $\$2 \times 3 = \$6$.

4. 50% of $200 = \dfrac{1}{2} \times 200 = 100$ items.

5. Unit rate: $\$24 \div 3 = \$8$ per shirt. 10 shirts: $10 \times \$8 = \$80$.

6. $\dfrac{3}{5} \times \dfrac{5}{1} = \dfrac{15}{5} = 3$.

7. $73 \div 24 = 3$ R1; bring down $4 \to 14 \div 24 = 0$ R14; bring down $4 \to 144 \div 24 = 6$. Answer: 306. The zero in the tens place must not be skipped. Check: $306 \times 24 = 7{,}344$.

Find more at
ViewMath.com/AK-Grade6

ViewMath.com

8 Multiply as whole numbers: $12 \times 3 = 36$. Count decimal places: 1.2 has 1 and 0.3 has 1, total 2. Place the decimal: 0.36.

9 $|6| = 6$ and $|-6| = 6$. Both 6 and -6 are 6 units from zero, so both have an absolute value of 6.

10 In Quadrant II, the x-coordinate is negative (left of the y-axis) and the y-coordinate is positive (above the x-axis), giving the sign convention $(-, +)$.

11 Giving away 6 means subtracting: $x - 6$.

12 $4(a + b)$ is a single term — it is one product. (If you distribute it to $4a + 4b$, it would be 2 terms.)

13 $3^2 + 3 = 9 + 3 = 12$.

14 $8 \times y + 8 \times 3 = 8y + 24$

15 $d = 60t$ means distance equals 60 miles per hour times the number of hours.

16 Subtract 24: $x = 50 - 24 = 26$.

17 "At most" means the value can equal the number or be less, so it should be $\leq$, not $<$.

18 "More than 75" is $t > 75$: open circle (not including 75), shade right (greater values).

19 2 cups per batch, b batches: $f = 2b$.

20 $A = \frac{1}{2} \times 5 \times 12 = 30 \ in^2$.

Find more at
ViewMath.com/AK-Grade6

21 Vertical segment: $|5 - (-2)| = |7| = 7$ units.

22 Base $= 10 - 2 = 8$. Height $= 8 - 1 = 7$. Area $= \frac{1}{2} \times 8 \times 7 = 28$ square units.

23 $SA = 2(5)(3) + 2(5)(4) + 2(3)(4) = 30 + 40 + 24 = 94$ cm^2. Yes, it is correct.

24 There are always 3 feet in one yard. This is not a statistical question because the answer does not vary.

25 Current mean $= (3 + 5 + 4 + 6 + 22) \div 5 = 40 \div 5 = 8$. Adding 5: new mean $= 45 \div 6 = 7.5$. Since $5 < 8$, the mean decreases.

26 Distances from 60: $10, 5, 0, 5, 10$. $MAD = (10 + 5 + 0 + 5 + 10) \div 5 = 30 \div 5 = 6$.

27 A dot plot has one dot per data value, so 12 values produce 12 dots. A histogram may have fewer bars (since values are grouped into intervals).

28 $IQR = Q3 - Q1 = 40 - 20 = 20$.

29 Team 1 range $= 68 - 60 = 8$. Team 2 range $= 72 - 58 = 14$. Team 2 has the greater range.

30 "At least 2" means 2 or more. Students with 2 pets: 5; with 3 pets: 4. Total: $5 + 4 = 9$.

✅ Practice Test 4 — Answer Key

1 B   **2** 3.5 hours   **3** C   **4** 87.5%; $\dfrac{7}{8}$ **5** B   **6** C   **7** B   **8** 1.44

9 C   **10** B   **11** A   **12** 7 and 4   **13** B   **14** $11m + 2$ **15** B   **16** A   **17** D

Find more at
ViewMath.com/AK-Grade6

 **18** B    **19** C    **20** C    **21** B    **22** B    **23** A

 **24** The teacher has one specific age, so the answer does not vary.    **25** B    **26** C    **27** B   **28** B

 **29** Fertilizer A: min $= 8$, $Q_1 = 15$, med $= 24$, $Q_3 = 32$, max $= 42$. Range $= 34$, IQR $= 17$. Fertilizer B: min $= 18$, Q_1

30 B

💡 Time to Learn! 💡

Review the explanations below, **especially for the questions you missed.**

Understanding why each answer is correct builds stronger problem-solving skills.

Tip: Circle any questions you got wrong, then read their explanation carefully.

📖 Practice Test 4 — Detailed Explanations

1 "3 pencils for every 1 eraser" means pencils to erasers is $3 : 1$.

2 $49 \div 14 = 3.5$ hours.

3 The ratio is $4 : 10 = 2 : 5$. Row 3 should be $12 : 30$ (since $4 \times 3 = 12$ and $10 \times 3 = 30$), but it shows $12 : 25$.

4 $0.875 \times 100 = 87.5\%$. $0.875 = \dfrac{875}{1000} = \dfrac{7}{8}$.

5 Unit rate: $180 \div 3 = 60$ mph. In 7 hours: $60 \times 7 = 420$ miles.

6 $\dfrac{2}{3} \times \dfrac{5}{4} = \dfrac{10}{12} = \dfrac{5}{6}$.

Find more at
ViewMath.com/AK-Grade6

7. The four steps of the standard algorithm are **Divide, Multiply, Subtract, Bring Down**. After dividing, you multiply the partial quotient by the divisor before subtracting.

8. Multiply as whole numbers: $36 \times 4 = 144$. Count decimal places: $1 + 1 = 2$. Place the decimal: 1.44. Check: $3.6 \times 0.4 = 1.44$.

9. Distance from sea level is measured by absolute value. $|-40| = 40$ and $|25| = 25$. Since $40 > 25$, the diver is farther from sea level. The sign tells direction (above or below), not distance.

10. Quadrant II has sign convention $(-, +)$. Since $-4 < 0$ and $7 > 0$, the point $(-4, 7)$ is in Quadrant II. A common mistake is choosing Quadrant III, which has both coordinates negative.

11. $\frac{n}{3}$ is "n divided by 3." Adding 7 gives "7 more than n divided by 3."

12. 7 and 4 are the terms without variables, so they are constants.

13. $P = 2(7) + 2(3) = 14 + 6 = 20$ cm.

14. Distribute: $8m + 2 + 3m$. Combine: $8m + 3m = 11m$. Result: $11m + 2$.

15. A square has 4 equal sides, so its perimeter is 4 times the side length: $P = 4s$.

16. Start at unknown x, add 8, arrive at 14: $x + 8 = 14$, so $x = 6$.

17. $2.5 \geq 3$ is false. $2.5 < 3$, so it is not a solution.

18. $\leq$ means 5 is included (closed circle). Less than 5 is to the left (shade left).

19. $i = 12(5) = 60$ inches.

Find more at
ViewMath.com/AK-Grade6

20 $A = \frac{1}{2} \times 20 \times 7 = 70\ in^2$.

21 Horizontal side: $|7 - 1| = 6$. Vertical side: $|2 - (-4)| = 6$. Both sides are 6 units, so it is a square. The side length is 6.

22 Length $= |5 - (-1)| = 6$. Width $= |3 - (-2)| = 5$. Area $= 6 \times 5 = 30$ square units.

23 Net A is a valid cross-shaped cube net. Net B has two squares on the same side, which causes overlap. Net C forms a 2×3 block, which doesn't fold into a cube.

24 A statistical question expects data that varies. The teacher's age is a single fixed number.

25 Ordered: $6, 7, 9, 11, 12$. The middle (3rd) value is 9.

26 IQR measures the spread of the middle 50%. A smaller IQR (4) means those values are closer together.

27 Histogram bins cover consecutive ranges with no overlap and no numbers skipped, so the bars sit right next to each other.

28 A long right whisker means some data values extend far above Q3, indicating the data is skewed to the right.

29 Fertilizer B's higher median shows its typical plant is taller. Its smaller IQR and range show the plant heights are less spread out. Fertilizer B is the better choice if you want tall, consistent growth.

30 A steep upward line means a sharp increase. A flat line means no change. So the value increased sharply then stayed the same.

☑ Practice Test 5 — Answer Key

Find more at
ViewMath.com/AK-Grade6

 10    C  C  D  C 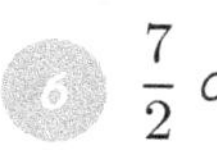 $\frac{7}{2}$ or $3\frac{1}{2}$  C  C  B

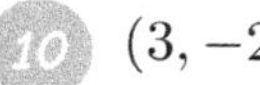 $(3, -2)$  $z - 9$  D  25    B  B  $w = 72$ 17 C 18 C

19 C 20 B 21 20 units   22 B 23 A 24 B 25 B 26 B 27 25

28 B 29 Group X: median = \$10, range = \$15, IQR = \$7. Group Y: median = \$11, range = \$3, IQR = \$2.

30 B

💡 Time to Learn! 💡

Review the explanations below, **especially for the questions you missed**.

Understanding why each answer is correct builds stronger problem-solving skills.

Tip: Circle any questions you got wrong, then read their explanation carefully.

📖 Practice Test 5 — Detailed Explanations

1. Total parts $= 3 + 2 = 5$. Each part $= 25 \div 5 = 5$. Swimming $= 2 \times 5 = 10$.

2. $240 \div 8 = 30$ miles per gallon.

3. $4 \times 2.5 = 10$ loaves, so $6 \times 2.5 = 15$ cups. Alternatively, unit rate: $6 \div 4 = 1.5$ cups per loaf, then $1.5 \times 10 = 15$.

4. Zoo $= 60\%$. Remaining $= 40\%$. Farm is twice Park, so Farm $= 2x$ and Park $= x$, giving $3x = 40\%$, $x \approx 13.3\%$. Farm $\approx 26.7\% \approx 27\%$.

5. $8 \times 5 = 40$ km.

Find more at
ViewMath.com/AK-Grade6

6 $\dfrac{7}{8} \div \dfrac{1}{4} = \dfrac{7}{8} \times \dfrac{4}{1} = \dfrac{28}{8} = \dfrac{7}{2} = 3\dfrac{1}{2}$ *pieces.*

7 $15 \div 5 = 3$; *bring down* $7 \to 7 \div 5 = 1$ *R2; bring down* $5 \to 25 \div 5 = 5$. *Answer:* 315. *Check:* $315 \times 5 = 1{,}575$.

8 0.6 *has 1 decimal place and* 0.7 *has 1 decimal place. Add them:* $1 + 1 = 2$ *decimal places. The product is* 0.42.

9 $|-10| = 10$ *because* -10 *is 10 units from zero. Absolute value is never negative. A common mistake is to think* $|-10| = -10$, *but distance from zero is always positive or zero.*

10   *Start at* $(0, 0)$. *Move 4 left and 3 up:* $(-4, 3)$. *Then move 7 right:* $-4 + 7 = 3$, *and 5 down:* $3 - 5 = -2$. *The final point is* $(3, -2)$.

11   *"9 fewer than* z*" means subtract 9 from* z: $z - 9$.

12 $6(n + 2)$ *means* $6 \times (n + 2)$. *The two factors being multiplied are 6 and* $(n + 2)$.

13 $4^2 + 2(4) + 1 = 16 + 8 + 1 = 25$.

14   *Distribute:* $4 \times 3y - 4 \times 2 = 12y - 8$.

15   *Notebooks cost* $3n$ *and pens cost* $1 \times p = p$. *Total:* $3n + p$.

16   *Multiply by 8:* $w = 9 \times 8 = 72$.

17 $t \le 30$ *means 30 or less, which is "no more than 30 degrees."*

18   *Closed circle means 3 is included* $(\ge$ *or* $\le)$. *Shading right means greater values:* $x \ge 3$.

Find more at
ViewMath.com/AK-Grade6

19   You choose x (independent), and y is computed from it (dependent). y increases as x increases.

20 $P = \frac{1}{2} \times 10 \times 6 = 30$. $Q = \frac{1}{2} \times 10 \times 8 = 40$. Difference $= 40 - 30 = 10\ cm^2$.

21   Each side $= 5$. Perimeter $= 4 \times 5 = 20$ units.

22   Base $= |8 - 2| = 6$. Height $= |7 - 1| = 6$. Area $= \frac{1}{2} \times 6 \times 6 = 18$ square units.

23 $SA = 2(5)(5) + 2(5)(10) + 2(5)(10) = 50 + 100 + 100 = 250\ in^2$.

24   Different sixth graders spend different amounts of time on homework, so the answers vary. The other questions all have a single fixed answer.

25   Mean $= (4 + 6 + 5 + 7 + 3) \div 5 = 25 \div 5 = 5$.

26   Distances from 24: $4, 2, 0, 2, 4$. MAD $= (4 + 2 + 0 + 2 + 4) \div 5 = 12 \div 5 = 2.4$.

27 $5 + 10 + 8 + 2 = 25$ students.

28   If the median is closer to $Q1$, the distance from median to $Q3$ is larger, meaning the upper half of the middle 50% is more spread out.

29   Group X: $Q_1 = 8$, $Q_3 = 15$, IQR $= 7$. Group Y: $Q_1 = 10$, $Q_3 = 12$, IQR $= 2$. Group Y gets a slightly higher typical allowance ($11 vs. $10) and allowances are much more consistent (IQR 2 vs. 7, range 3 vs. 15).

30   The axes show sports (categories) and number of students (counts). The best title describes a survey of students' favorite sports.

✅ Practice Test 6 — Answer Key

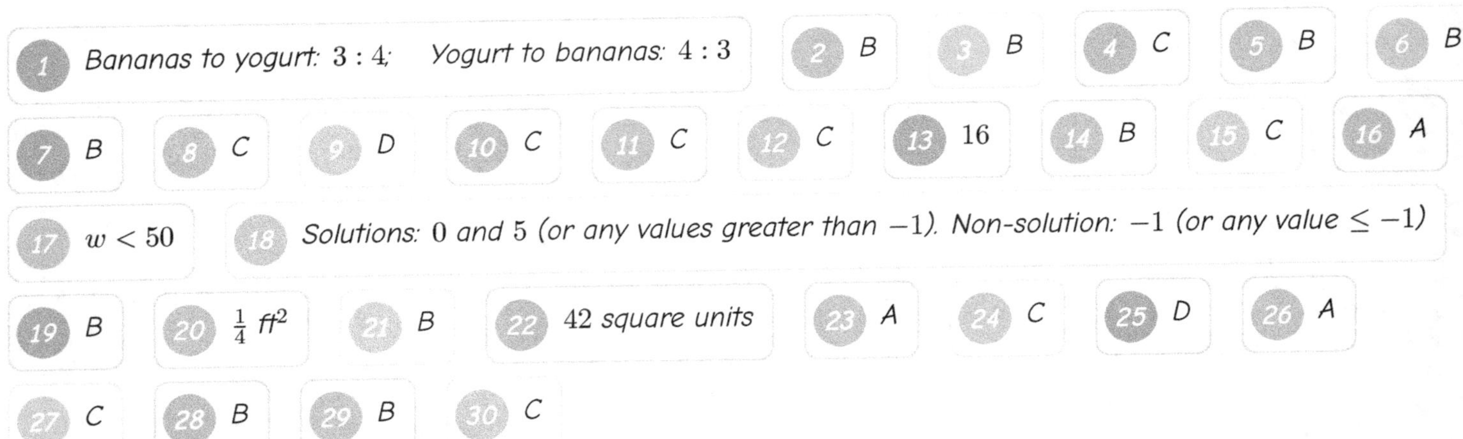

1 Bananas to yogurt: $3 : 4$;   Yogurt to bananas: $4 : 3$ **2** B   **3** B   **4** C   **5** B   **6** B

7 B   **8** C   **9** D   **10** C   **11** C   **12** C   **13** 16   **14** B   **15** C   **16** A

17 $w < 50$ **18** Solutions: 0 and 5 (or any values greater than -1). Non-solution: -1 (or any value ≤ -1)

19 B   **20** $\frac{1}{4}$ ft^2 **21** B   **22** 42 square units   **23** A   **24** C   **25** D   **26** A

27 C   **28** B   **29** B   **30** C

💡 Time to Learn! 💡

Review the explanations below, **especially for the questions you missed**.

Understanding why each answer is correct builds stronger problem-solving skills.

Tip: Circle any questions you got wrong, then read their explanation carefully.

📖 Practice Test 6 — Detailed Explanations

1 "3 bananas for every 4 cups of yogurt" gives $3 : 4$. Flip the order for yogurt to bananas: $4 : 3$.

2 Brand X: $\$8.75 \div 5 = \$1.75$ each. Brand Y: $\$4.50 \div 3 = \$1.50$ each. Brand Y is cheaper.

3 $2 \times 8 = 16$ students, so $3 \times 8 = 24$ pencils.

4 Multiply by 100: $0.35 \times 100 = 35\%$.

Find more at
ViewMath.com/AK-Grade6

5　*"How much for one" is a unit rate problem. Divide to find the amount per 1 unit.*

6　*The reciprocal of a fraction is found by swapping the numerator and denominator:* $\frac{3}{7} \to \frac{7}{3}$.

7　$2{,}016 \div 14 = 144$. *Check:* $144 \times 14 = 2{,}016$.

8　*Move the decimal one place in both numbers:* $93.6 \div 12 = 7.8$. *Check:* $7.8 \times 1.2 = 9.36$.

9　*The opposite of* -3 *is 3. Both* -3 *and 3 are 3 units from zero, but on opposite sides of the number line.*

10　*The x-coordinate -3 tells you to move 3 units to the left, and the y-coordinate 4 tells you to move 4 units up. Choice D swaps the coordinates.*

11　*Half of n is $\frac{n}{2}$. "6 more" means add 6:* $\frac{n}{2} + 6$.

12　*Part C points to 3, which is a term with no variable — a constant.*

13　$8^2 = 64$. *Then* $64 \div 4 = 16$.

14　*Distribute:* $3x + 12 + 2x$. *Combine:* $3x + 2x = 5x$. *Result:* $5x + 12$.

15　*Hourly earnings:* $8h$. *Plus tip:* $8h + 15$.

16　*Subtract 3.5:* $x = 10 - 3.5 = 6.5$.

17　*"Less than 50" means strictly under 50:* $w < 50$.

18　*Any number greater than -1 is a solution. -1 itself is not, since $>$ does not include the boundary.*

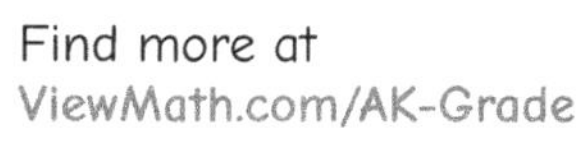

Find more at
ViewMath.com/AK-Grade6

ViewMath.com

19 The independent variable (number of cars) goes on the x-axis.

20 $A = \frac{1}{2} \times \frac{3}{4} \times \frac{2}{3} = \frac{1}{2} \times \frac{6}{12} = \frac{1}{2} \times \frac{1}{2} = \frac{1}{4}\ ft^2.$

21 $(5, -1)$ and $(5, 4)$ share the same x-coordinate, so the segment is vertical.

22 Length $= |9 - 2| = 7$. Width $= |5 - (-1)| = 6$. Area $= 7 \times 6 = 42$ square units.

23 $8 \times 5 \times 2 = 80$ is the volume. The actual surface area is $2(40) + 2(16) + 2(10) = 132\ cm^2$.

24 Exercise time varies from student to student. The other questions each have one fixed answer.

25 Mean $= sum \div 4 = 25$, so sum $= 25 \times 4 = 100$.

26 Team A range: $68 - 60 = 8$. Team B range: $88 - 40 = 48$. Team A is much more consistent.

27 There are no data values between 25 and 30, creating a gap in the data.

28 Lower half: $3, 5, 7$. The median of the lower half is 5, so $Q1 = 5$.

29 A small IQR means the middle 50% is tightly packed. A large range means the minimum and maximum are far apart, indicating extreme values stretch the data.

30 Add all frequencies: $5 + 8 + 3 + 4 = 20$ students.

📋 Practice Test 7 — Answer Key

Find more at
ViewMath.com/AK-Grade6

ViewMath.com

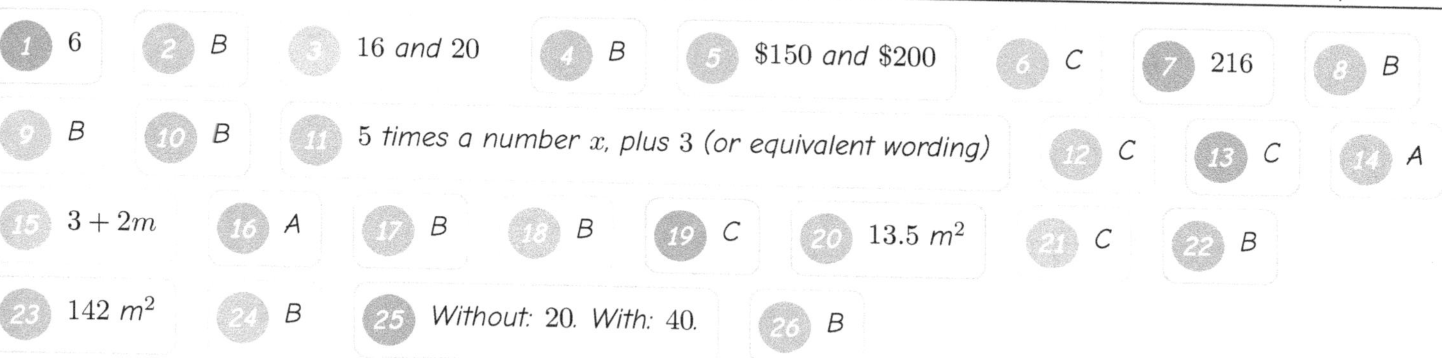

1. 6
2. B
3. 16 *and* 20
4. B
5. $150 *and* $200
6. C
7. 216
8. B
9. B
10. B
11. 5 *times a number* x, *plus* 3 (*or equivalent wording*)
12. C
13. C
14. A
15. $3 + 2m$
16. A
17. B
18. B
19. C
20. 13.5 m^2
21. C
22. B
23. 142 m^2
24. B
25. *Without:* 20. *With:* 40.
26. B
27. 18 *students; median* $= 8$; *skewed left with peak at* 8.
28. B
29. C
30. B

💡 Time to Learn! 💡

Review the explanations below, **especially for the questions you missed.**

Understanding why each answer is correct builds stronger problem-solving skills.

Tip: *Circle any questions you got wrong, then read their explanation carefully.*

📖 Practice Test 7 — Detailed Explanations

1. $21 \div 7 = 3$, *so multiply both by* 3: *iced tea* $= 2 \times 3 = 6$.

2. $150 \div 5 = 30$ *gallons per hour.*

3. *Row 2:* $5 \times 2 = 10$, *so* $8 \times 2 = 16$. *Row 3:* $8 \times 4 = 32$, *so* $5 \times 4 = 20$.

4. 70% *of* $30 = 0.70 \times 30 = 21$ *students.*

5. *Total parts* $= 7$. *Each part* $= \$350 \div 7 = \$50$. *Family 1:* $3 \times \$50 = \$150$. *Family 2:* $4 \times \$50 = \$200$.

6 Keep, Change, Flip: $\frac{1}{2} \times \frac{4}{1} = \frac{4}{2} = 2$.

7 $90 \div 42 = 2$ R6; bring down 7 $\rightarrow$ $67 \div 42 = 1$ R25; bring down 2 $\rightarrow$ $252 \div 42 = 6$. Answer: 216. Check: $216 \times 42 = 9{,}072$.

8 Multiply: 2.75×3. As whole numbers: $275 \times 3 = 825$. Two decimal places: $\$8.25$.

9 The opposite of a number is the same distance from zero on the other side of the number line. The opposite of 5 is -5 because both are 5 units from zero.

10 When the y-coordinate is 0, the point lies on the x-axis. The point $(-6, 0)$ is 6 units to the left of the origin on the x-axis. A common mistake is thinking a negative x-value means the point is on the y-axis.

11 $5x$ means "5 times x" and $+3$ means "plus 3."

12 $5m - 2n + 7 + m$ has 4 terms: $5m$, $2n$, 7, and m.

13 $2(9) + 2(4) = 18 + 8 = 26$.

14 Like terms: $5p + 2p + 4p = 11p$. Constants: $3 - 1 = 2$. Result: $11p + 2$.

15 Flat fee: $\$3$. Per-mile cost: $2m$. Total: $3 + 2m$.

16 Divide both sides by 5: $n = 45 \div 5 = 9$.

17 "Greater than 7" uses the $>$ symbol: $x > 7$.

18 $x > 2$ does NOT include 2, so the circle should be open, not closed.

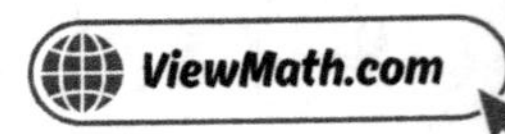

19 Time (hours) is independent — it passes regardless. Battery level depends on time.

20 $A = \frac{1}{2} \times 4.5 \times 6 = 13.5\ m^2$.

21 The base goes from $(0, 0)$ to $(8, 0)$. Distance $= |8 - 0| = 8$ units.

22 Side $= |4 - (-2)| = 6$. Area $= 6 \times 6 = 36$ square units.

23 $SA = 2(7)(5) + 2(7)(3) + 2(5)(3) = 70 + 42 + 30 = 142\ m^2$.

24 The dot plot shows data ranging from 0 to 5, with different students giving different answers. This variability confirms the question is statistical.

25 Without: $(15 + 20 + 25) \div 3 = 20$. With: $(15 + 20 + 25 + 100) \div 4 = 160 \div 4 = 40$. The 100 doubles the mean.

26 MAD (Mean Absolute Deviation) is calculated by finding the average of the distances of each data value from the mean.

27 Total: $1 + 2 + 4 + 6 + 3 + 2 = 18$. Median is average of 9th and 10th values. Counting from 5: $1, 3, 7$ (past 9th). Both the 9th and 10th values are 8. Median $= 8$. Data peaks at 8 with a tail to the left (skewed left).

28 The five-number summary consists of minimum, first quartile (Q1), median, third quartile (Q3), and maximum.

29 One extreme outlier pulls the mean away from where most of the data lies. The median would be a better choice.

30 Total: $24 + 30 = 54$. Fraction of boys: $\dfrac{24}{54} = \dfrac{4}{9}$.

Find more at
ViewMath.com/AK-Grade6

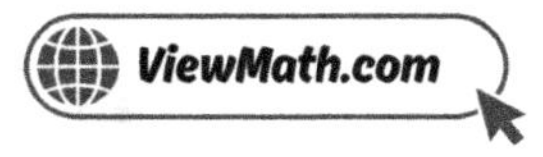

✅ Practice Test 8 — Answer Key

1 B

2 Part A: 0.5 laps per minute (or 1 lap every 2 minutes).   Part B: 7 laps.

3 15

4 C

5 40 minutes

6 D

7 A

8 B

9 C

10 Part A: Quadrant II;   Part B: $(8, 1)$, Quadrant I;   Part C: $(-8, -1)$, Quadrant III

11 B

12 B

13 $23

14 $6x + 12$

15 $n + 12 - 5$ or $n + 7$

16 $a = 7$

17 $n = 4$ is NOT a solution to $n > 4$ (since 4 is not greater than 4). $n = 4$ IS a solution to $n \geq 4$ (since 4 equals 4).

18 C

19 $w = 45 - 3t$; empty when $t = 15$ minutes

20 A

21 $(9, -5)$

22 B

23 B

24 Answers vary. Example: "How many glasses of water do you drink each day?"

25 13

26 C

27 5000

28 C

29 Not always.

30 D

💡 Time to Learn! 💡

Review the explanations below, **especially for the questions you missed**.

Understanding why each answer is correct builds stronger problem-solving skills.

Tip: Circle any questions you got wrong, then read their explanation carefully.

📖 Practice Test 8 — Detailed Explanations

1 Dogs have 4 parts. Each part = 2 animals. Dogs = $4 \times 2 = 8$.

Find more at
ViewMath.com/AK-Grade6

2   Part A: From the graph, the swimmer completes 1 lap every 2 minutes, so the rate is $\frac{1}{2}$ lap per minute. Part B: $14 \div 2 = 7$ laps.

3 $2 \times 5 = 10$ strawberries, so $3 \times 5 = 15$ cups of yogurt.

4 $100\% = \frac{100}{100} = 1$ whole.

5 $1{,}000 \div 250 = 4$ batches. $10 \times 4 = 40$ minutes.

6   The student flipped the first fraction ($\frac{4}{5} \to \frac{5}{4}$) instead of the second ($\frac{2}{3} \to \frac{3}{2}$). The correct answer is $\frac{4}{5} \times \frac{3}{2} = \frac{12}{10} = \frac{6}{5}$.

7   To check a division, multiply the quotient by the divisor. If $4{,}752 \div 12 = 396$, then 396×12 should equal $4{,}752$.

8   Multiply as whole numbers: $12 \times 5 = 60$. Count decimal places: 0.12 has 2 and 0.5 has 1, total 3. Place the decimal: $0.060 = 0.06$.

9   The opposite of 0 is 0 itself. Zero is neither positive nor negative and sits right in the middle of the number line. It is the only number that is its own opposite.

10   Part A: $(-8, 1)$ has signs $(-, +)$, which is Quadrant II. Part B: Reflecting across the y-axis flips the x-sign: $(-8, 1) \to (8, 1)$. Signs $(+, +)$ is Quadrant I. Part C: Reflecting across the x-axis flips the y-sign: $(-8, 1) \to (-8, -1)$. Signs $(-, -)$ is Quadrant III.

11   "8 less than a number" means subtract 8 from n: $n - 8$.

12   The term d means $1 \cdot d$, so the coefficient is 1. (The subtraction sign belongs to the operation, not the coefficient of d itself in the original expression.)

Find more at
ViewMath.com/AK-Grade6

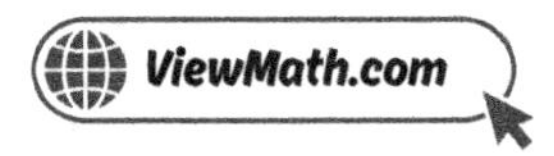

13 $5 + 3(6) = 5 + 18 = 23$ *dollars.*

14 *Perimeter* $= 2(2x + 5) + 2(x + 1) = 4x + 10 + 2x + 2 = 6x + 12.$

15 *Start with* n, *add* 12, *subtract* 5: $n + 12 - 5 = n + 7.$

16 $a = 105 \div 15 = 7.$

17 $>$ *does not include equality.* $\geq$ *includes equality.*

18 $-3 \geq -3$ *is true (they are equal). The other values are less than* $-3.$

19 *Set* $w = 0$: $0 = 45 - 3t$, *so* $3t = 45$, $t = 15$ *minutes.*

20 *Triangle A:* $\frac{1}{2} \times 4 \times 4 = 8.$ *Triangle B:* $\frac{1}{2} \times 5 \times 3 = 7.5.$ *Triangle A has the greater area.*

21 *The fourth vertex must share an* x-*value with* $(9, 3)$ *and a* y-*value with* $(2, -5)$, *giving* $(9, -5).$

22 *Base* $= 8$, *height* $= 6.$ *Area* $= \frac{1}{2} \times 8 \times 6 = 24$ *square units.*

23 *A net is a flat pattern that, when folded, forms a 3D shape. Each section of the net becomes a face.*

24 *Any question about food where different classmates would give different answers is acceptable.*

25 *Need total* $= 13 \times 5 = 65.$ *Current total* $= 10 + 14 + 12 + 16 = 52.$ *Fifth value* $= 65 - 52 = 13.$

26 *The range uses only the maximum and minimum values, so one extreme outlier can make the range very large. The IQR ignores the most extreme values.*

Find more at
ViewMath.com/AK-Grade6

27 5000 has 6 dots, the highest frequency. It is the mode.

28 A larger IQR means the middle 50% of the data covers a wider range. Class B's middle scores are more spread out.

29 A single outlier can make the range very large while the rest of the data is tightly clustered. The IQR gives a better picture of overall spread. For example, $\{1, 50, 51, 52, 53\}$ has range 52 but most values are close together.

30 Red: 12, Yellow: 4. Difference: $12 - 4 = 8$.

☑ Practice Test 9 — Answer Key

1 C   **2** B   **3** B   **4** C   **5** C   **6** A   **7** B

8 Part A: 27.54 km;   Part B: 9.18 km per day   **9** D

10 Part A: $A = (2, 5)$, $B = (5, 1)$, $C = (2, 1)$;   Part B: $A' = (-2, 5)$, $B' = (-5, 1)$, $C' = (-2, 1)$;   Part C: Quadrant II

11 B   **12** B   **13** A   **14** A   **15** C   **16** C   **17** C   **18** B   **19** C   **20** C

21 A   **22** C   **23** B   **24** C   **25** B   **26** C   **27** B   **28** B   **29** B

30 Monday, 0 more (they are equal at 35)

💡 Time to Learn! 💡

Review the explanations below, **especially for the questions you missed**.

Understanding why each answer is correct builds stronger problem-solving skills.

Tip: Circle any questions you got wrong, then read their explanation carefully.

Find more at
ViewMath.com/AK-Grade6

📖 Practice Test 9 — Detailed Explanations

1. Total parts $= 1 + 3 + 2 = 6$. Each part $= 18 \div 6 = 3$. Blue $= 3 \times 3 = 9$.

2. Divide: $120 \div 4 = 30$ pages per minute.

3. $3 \times 2 = 6$, so $5 \times 2 = 10$.

4. 88 out of $100 = 88\%$. When the total is 100, the number correct IS the percent.

5. $3 \times 4 = 12$ raisins, so $4 \times 4 = 16$ cups of granola.

6. $\dfrac{3}{4} \div \dfrac{1}{8} = \dfrac{3}{4} \times \dfrac{8}{1} = \dfrac{24}{4} = 6$ pieces.

7. $26 \div 16 = 1$ R10; bring down 8 $\to 108 \div 16 = 6$ R12; bring down 8 $\to 128 \div 16 = 8$. Answer: 168. Check: $168 \times 16 = 2{,}688$.

8. Part A: $8.76 + 12.30 + 6.48 = 27.54$ km. Part B: $27.54 \div 3 = 9.18$ km per day. Check: $9.18 \times 3 = 27.54$.

9. A number and its opposite are the same distance from zero on different sides, so they cancel each other out. For example, $6 + (-6) = 0$ and $-9 + 9 = 0$.

10. Part A: Reading from the grid, $A = (2, 5)$, $B = (5, 1)$, $C = (2, 1)$. Part B: Reflecting across the y-axis changes the sign of each x-coordinate: $A' = (-2, 5)$, $B' = (-5, 1)$, $C' = (-2, 1)$. Part C: All reflected vertices have negative x-coordinates and positive y-coordinates, which is the sign convention $(-, +)$ for Quadrant II.

11. The product of 4 and t is $4t$. The sum of 9 and that product is $9 + 4t$.

Find more at
ViewMath.com/AK-Grade6

🌐 ViewMath.com

12 The terms are $4x$, 9, and $2y$. Terms are separated by $+$ or $-$ signs.

13 $6(4) - 4^2 = 24 - 16 = 8$.

14 $7n$ and $4n$ are like terms. Add coefficients: $7 + 4 = 11$, so $7n + 4n = 11n$.

15 s represents the side length of any square. It can be any positive number.

16 Multiply by 6: $t = 5 \times 6 = 30$.

17 $5 > 5$ is false. 5 equals 5, but is not greater than 5. It would be a solution to $x \geq 5$.

18 Open circle means -2 is NOT included ($<$ or $>$). Shading left means less than: $x < -2$.

19 $c = 7(6) = 42$ dollars.

20 $A = \frac{1}{2} \times \frac{1}{2} \times \frac{1}{4} = \frac{1}{16}$ ft^2.

21 From $(-2, -2)$, moving 5 right gives $x = 3$, and 5 up gives $y = 3$. The opposite vertex is $(3, 3)$.

22 Length $= 6$, width $= 4$. Area $= 6 \times 4 = 24$ square units.

23 $6s^2 = 96$, so $s^2 = 16$, giving $s = 4$ cm.

24 $12 \times 9 = 108$ is a single fixed answer. It does not vary from person to person.

25 10 values. Median is average of 5th and 6th. Counting: 0(1), 2(3), 4(6). The 5th value is 4 and the 6th is 4. Median $= 4$.

Find more at
ViewMath.com/AK-Grade6

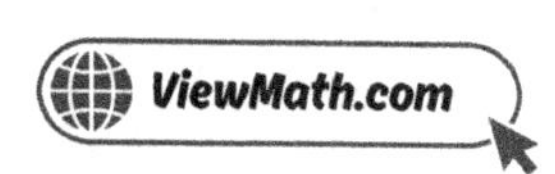

26   Adding 5 to every value shifts the max and min by the same amount, so the difference (range) stays unchanged.

27   Total runners $= 4 + 8 + 6 + 3 + 1 = 22$. The 20–29 group has 8 runners. Fraction $= \frac{8}{22}$.

28   Even number of values. The two middle values are 6 and 8. Median $= (6 + 8) \div 2 = 7$.

29 X: $Q_1 = 20$, $Q_3 = 60$, $IQR = 40$. Y: $Q_1 = 37$, $Q_3 = 43$, $IQR = 6$. Y is much smaller.

30   Monday: $15 + 20 = 35$. Tuesday: $25 + 10 = 35$. Both days had the same total of 35 fruits.

☑ Practice Test 10 — Answer Key

1 C	**2** B	**3** B

1 C   **2** B   **3** B   **4** A   **5** C   **6** $\frac{2}{3} \div \frac{1}{6} = 4$ **7** 252   **8** 21.6   **9** D

10 D   **11** $4x + 5$ **12** B   **13** $A = \$14$, $B = \$23$ **14** A   **15** B   **16** $m = 13$ **17** C

18 $x \leq 4$ **19** C   **20** B   **21** C   **22** B   **23** $247.2 \ cm^2$ **24** C   **25** B

26 Data set B   **27** B   **28** B   **29** C   **30** C

💡 Time to Learn! 💡

Review the explanations below, **especially for the questions you missed**.

Understanding why each answer is correct builds stronger problem-solving skills.

Tip: Circle any questions you got wrong, then read their explanation carefully.

Find more at
ViewMath.com/AK-Grade6

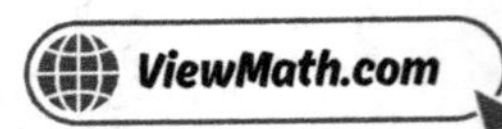

📖 *Practice Test 10 — Detailed Explanations*

1. Apple has 3 parts, orange has 5 parts. Total parts $= 8$. Total ounces $= 8 \times 4 = 32$.

2. Store A: $\$4.50 \div 10 = \$0.45$. Store B: $\$3.20 \div 8 = \$0.40$. Store B is cheaper per pencil.

3. $2 : 3$ multiplied by 3 gives $6 : 9$. The other pairs do not simplify to $2 : 3$.

4. Convert all to decimals: $30\% = 0.30$, $\dfrac{1}{3} \approx 0.333$, 0.35. Order: $0.30 < 0.333 < 0.35$.

5. $42 \div 6 = 7$ sections, which needs $7 + 1 = 8$ posts (one at each end). Note: if counting "per 6 feet" as a rate, $42 \div 6 = 7$, but fences need a post at the start too, giving 8.

6. $\dfrac{2}{3} \div \dfrac{1}{6} = \dfrac{2}{3} \times \dfrac{6}{1} = \dfrac{12}{3} = 4$. The shaded part of Bar B fits into the shaded part of Bar A exactly 4 times.

7. $37 \div 15 = 2$ R7; bring down $8 \to 78 \div 15 = 5$ R3; bring down $0 \to 30 \div 15 = 2$. Answer: 252. Check: $252 \times 15 = 3{,}780$.

8. Move the decimal one place in both numbers: $86.4 \div 4 = 21.6$. Check: $21.6 \times 0.4 = 8.64$.

9. The absolute value of 0 is 0 because zero is 0 units from itself. Zero is the only number whose absolute value is 0.

10. Quadrant IV has sign convention $(+, -)$. Since $2 > 0$ and $-5 < 0$, the point $(2, -5)$ is in Quadrant IV. A common error is confusing Quadrant IV with Quadrant II which has signs $(-, +)$.

11. There are 4 triangles (each worth x) and 5 unit squares. The expression is $4x + 5$.

Find more at
ViewMath.com/AK-Grade6

12 $\frac{n}{5} = \frac{1}{5} \times n$. The factors are $\frac{1}{5}$ and n.

13 $h = 2$: $8 + 3(2) = 14$. $h = 5$: $8 + 3(5) = 23$.

14   Distribute: $7k - 21 + 5$. Combine: $-21 + 5 = -16$. Result: $7k - 16$.

15   In 4 days you read $4r$ pages. Remaining: $100 - 4r$.

16   Subtract 17: $m = 30 - 17 = 13$.

17   "More than \$25" means greater than 25: $d > 25$.

18   Closed circle at 4 (included) and shading to the left (less than): $x \leq 4$.

19   Each dog has 4 legs: $L = 4d$.

20 $24 = \frac{1}{2} \times 8 \times h$, so $24 = 4h$, which gives $h = 6$ ft.

21   Same x-coordinate, so it is a vertical distance. $|5 - (-3)| = |8| = 8$ units.

22   Bottom rectangle: $8 \times 3 = 24$. Left rectangle above: $4 \times (7 - 3) = 4 \times 4 = 16$. Total: 40. This matches option B.

23   2 triangles: $2 \times \frac{1}{2}(6)(5.2) = 31.2$. 3 rectangles: $3 \times 6 \times 12 = 216$. Total: $31.2 + 216 = 247.2$ cm^2.

24   If you ask 10 people how long they spend eating lunch, the answers vary. Following the flowchart, varying answers lead to "Statistical."

Find more at
ViewMath.com/AK-Grade6

25 Ordered: $8, 9, 10, 11, 12, 13, 14$. Median (4th of 7) $= 11$.

26 A larger MAD means values are farther from the mean on average. Data set B's MAD of 10 indicates more variability than A's MAD of 2.

27 A histogram groups numerical data into equal intervals (bins) with no gaps between consecutive bars. A bar graph has gaps and shows categorical data.

28 $Q1 = 30$, $Q3 = 70$. $IQR = 70 - 30 = 40$.

29 For symmetric data, the mean is a good center. For skewed data, the median better represents the typical value since it is not pulled by the tail.

30 The title tells us the data is about math quiz scores and specifically from Period 3, not the whole school or year.

Well done checking your answers!

Keep practicing to strengthen your skills.

Find more at
ViewMath.com/AK-Grade6

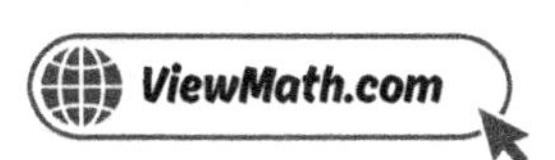

Author's Final Note

I hope you enjoyed this book as much as I enjoyed writing it. Whether you are a student working through the material, a parent supporting your child's learning, or a teacher guiding your class, I have tried to make this book as clear and engaging as possible. I hope I have succeeded. If you have any suggestions for improvement, please let me know. I would love to hear from you.

The accuracy of calculations is very important to me. We have done our best, but I also expect that I have made some minor errors. Constant improvement is the name of the game. If you find any errors, please let me know. I will fix them in the next edition.

For students: Your learning journey does not end here. I have written a series of books to help you learn math. Make sure you browse through them. I especially recommend workbooks and practice tests to help you prepare for your exams.

For parents: Thank you for investing in your child's education. I encourage you to explore the companion resources available online to help support your child outside the classroom.

For teachers: Thank you for the invaluable work you do every day. I hope this book serves as a useful resource in your classroom. Feel free to reach out if you have suggestions or would like to discuss how best to use this book with your students.

I also enjoy reading your reviews. If you have a moment, please leave a review on where you found this book. It will help others find this book. If you have any questions or comments, please feel free to contact me at drNazari@ViewMath.com.

And one last thing: Remember to use online resources for additional help. I recommend using the resources on `https://ViewMath.com` You can find video lessons, practice problems, and more. You can also use the online companion for this book to track your progress and access additional resources.

Wishing all students the best in their studies, parents every success in supporting their children, and teachers continued inspiration in their classrooms!

Dr. A. Nazari

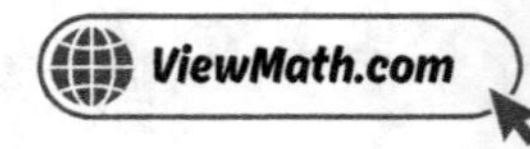

 *Great Job! Keep Learning with ViewMath!*

*Keep up the great work! Visit **viewmath.com/AK-Grade6** for free lessons, quizzes, and more.*

Study Guide

Workbook

Step-by-Step

3 Practice Tests

5 Practice Tests

7 Practice Tests

Find more at
ViewMath.com/AK-Grade6

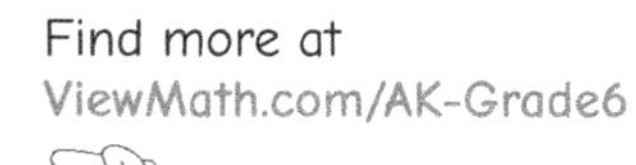

ViewMath.com